Practical Automated Machine Learning Using H2O.ai

Discover the power of automated machine learning,
from experimentation through to deployment to production

Salil Ajgaonkar

BIRMINGHAM—MUMBAI

Practical Automated Machine Learning Using H2O.ai

Publishing Product Manager: Dhruv Jagdish Kataria
Senior Editor: Nathanya Dias
Technical Editor: Devanshi Ayare
Copy Editor: Safis Editing
Project Coordinator: Farheen Fathima
Proofreader: Safis Editing
Indexer: Subalakshmi Govindhan
Production Designer: Ponraj Dhandapani
Marketing Coordinators: Shifa Ansari, Abeer Riyaz Dawe

First published: September 2022

Production reference: 1140922

Published by Packt Publishing Ltd.
Livery Place
35 Livery Street
Birmingham
B3 2PB, UK.

ISBN 978-1-80107-452-0

www.packt.com

Contributors

About the author

Salil Ajgaonkar is a software engineer experienced in building and scaling cloud-based microservices and productizing machine learning models. His background includes work in transaction systems, artificial intelligence, and cyber security. He is passionate about solving complex scaling problems, building machine learning pipelines, and data engineering. Salil earned his degree in IT from Xavier Institute of Engineering, Mumbai, India, in 2015 and later earned his master's degree in computer science from Trinity College Dublin, Ireland, in 2018, specializing in future networked systems. His work history includes the likes of BookMyShow, Genesys, and Vectra AI.

I would like to thank my lovely wife, Oshin, for supporting me and making sure I gave my best effort to writing this book. Also, thanks to my parents, who taught me to always say "yes" to all opportunities that come my way.

I would also like to thank the Packt team for giving me the opportunity to write this book and play my part in giving back to the programming community. Special thanks to Dhruv for coordinating the publication effort, Kirti for scheduling and keeping things on track over the year, and Nathanya and the editorial team for ensuring that the book is of the highest quality.

Last but not least, special thanks to Dr. Emir Muñoz for his valuable insights into the technical aspects of the book.

About the reviewer

Emir Muñoz is a senior manager of machine learning at Genesys, where he works on projects to enhance customer experience using artificial intelligence, machine learning, and data science. He has experience in academia and industry, which he uses to leverage emerging technologies and algorithms to deliver innovative solutions. Currently, he leads a team that mines contact center data to train machine learning models to optimize contact center routing.

Emir holds a PhD in computer science with a specialization in machine learning. He also received a BEng in informatics and an MSc in computer engineering. He is the author of several papers and patents on the topics of semantic web, machine learning, knowledge graphs, and contact center analytics.

Table of Contents

Part 2: H2O AutoML Deep Dive

3

Understanding Data Processing   79

4

Understanding H2O AutoML Architecture and Training   121

5

Understanding AutoML Algorithms   145

6

Understanding H2O AutoML Leaderboard and Other Performance Metrics   187

7

Working with Model Explainability   215

Part 3: H2O AutoML Advanced Implementation and Productization

8

Exploring Optional Parameters for H2O AutoML   247

9

Exploring Miscellaneous Features in H2O AutoML   263

10

Working with Plain Old Java Objects (POJOs)   279

11

Working with Model Object, Optimized (MOJO)   289

12

Working with H2O AutoML and Apache Spark   303

13

Using H2O AutoML with Other Technologies   331

Index   371

Other Books You May Enjoy   376

Preface

With the huge amount of data being generated over the internet and the benefits that Machine Learning (ML) predictions bring to businesses, ML implementation has become a low-hanging fruit that everyone is striving for. The complex mathematics behind it, however, can be discouraging for a lot of users. This is where H2O comes in – it automates various repetitive steps, and this encapsulation helps developers focus on results rather than handling complexities.

You'll begin by understanding how H2O's AutoML simplifies the implementation of ML by providing a simple, easy-to-use interface to train and use ML models. Next, you'll see how AutoML automates the entire process of training multiple models, optimizing their hyperparameters, as well as explaining their performance. As you advance, you'll find out how to leverage a Plain Old Java Object (POJO) and Model Object, Optimized (MOJO) to deploy your models to production. Throughout this book, you'll take a hands-on approach to implementation using H2O that'll enable you to set up your ML systems in no time.

By the end of this H2O book, you'll be able to train and use your ML models using H2O AutoML, right from experimentation all the way to production without a single need to understand complex statistics or data science.

Who this book is for

This book is for engineers and data scientists who want to quickly adopt ML into their products without worrying about the internal intricacies of training ML models.

If you are someone who wants to incorporate ML into your software system but doesn't know where to start or doesn't have much expertise in the domain of ML, then you will find this book useful.

What this book covers

Chapter 1, *Understanding H2O AutoML Basics*, talks about an AutoML technology by H2O.ai named H2O AutoML and implements a basic setup of the technology to see it in action.

Chapter 2, *Working with H2O Flow (H2O's Web UI)*, explores H2O's Web UI called H2O Flow and shows how we can set up our H2O AutoML system using the Web UI without writing a single line of code.

Chapter 3, *Understanding Data Processing*, explores some of the common data processing functionalities that we can perform using H2O's in-built dataframe manipulation operations.

Chapter 4, Understanding H2O AutoML Training and Architecture, deep dives into understanding the high-level architecture of H2O technology and teaches us how H2O AutoML trains all the models and optimizes their hyperparameters.

Chapter 5, Understanding AutoML Algorithms, explores the various ML algorithms that H2O AutoML uses to train various models.

Chapter 6, Understanding H2O AutoML Leaderboard and Other Performance Metrics, explores the different performance metrics that are used in the AutoML Leaderboard as well as some additional metrics that are important for users to know.

Chapter 7, Working with Model Explainability, explores the H2O explainability interface and helps us to understand the various explainability features that we get as outputs.

Chapter 8, Exploring Optional Parameters for H2O AutoML, looks at some of the optional parameters that are available to us when configuring our AutoML training and shows how we can use them.

Chapter 9, Exploring Miscellaneous Features in H2O AutoML, explores two unique features of H2O AutoML. The first one is H2O AutoML's compatibility with the scikit-learn library and the second one is H2O AutoML's inbuilt logging system for debugging AutoML training issues.

Chapter 10, Working with Plain Old Java Objects (POJOs), covers model POJOs and how we can extract and use them to make predictions in production environments.

Chapter 11, Working with Model Object, Optimized (MOJO), covers model MOJOs, how they are different from model POJOs, how to view them, and how we can extract and use them to make predictions in production environments.

Chapter 12, Working with H2O AutoML and Apache Spark, explores in detail how H2O AutoML can be used along with Apache Spark using H2O Sparkling Water.

Chapter 13, Using H2O AutoML with Other Technologies, explores how we can use H2O models in collaboration with other commonly used technologies in the ML domain, such as Spring Boot Web applications and Apache Storm.

To get the most out of this book

Basic knowledge of statistics and programming is beneficial. Some understanding of ML and Python will be helpful. You will need Python installed on your computer, preferably with version 3.7 or above, or R installed on your computer with version 4.0 or above. All code examples have been tested using Python 3.10 and R 4.1.2 on Windows 10 OS and Ubuntu 22.04.1 LTS. However, they should work with future version releases too.

Software/hardware covered in the book	Operating system requirements
Python 3.10	Windows, macOS, or Linux
R 4.1.2	
H2O 3.36.1.4	
Java 11	
Spark 3.2	
Scala 2.13	
Maven 3.8.6	

If you are using the digital version of this book, we advise you to type the code yourself or access the code from the book's GitHub repository (a link is available in the next section). Doing so will help you avoid any potential errors related to the copying and pasting of code.

Download the example code files

You can download the example code files for this book from GitHub at `https://github.com/PacktPublishing/Practical-Automated-Machine-Learning-on-H2O`. If there's an update to the code, it will be updated in the GitHub repository.

We also have other code bundles from our rich catalog of books and videos available at `https://github.com/PacktPublishing/`. Check them out!

Download the color images

We also provide a PDF file that has color images of the screenshots and diagrams used in this book. You can download it here: `https://packt.link/IighZ`.

Conventions used

There are a number of text conventions used throughout this book.

`Code in text`: Indicates code words in text, database table names, folder names, filenames, file extensions, pathnames, dummy URLs, user input, and Twitter handles. Here is an example: "The only dependency on using POJO models is the `h2o-genmodel.jar` file."

A block of code is set as follows:

```
import h2o
h2o.init()
```

When we wish to draw your attention to a particular part of a code block, the relevant lines or items are set in bold:

```
data_frame = h2o.import_file("Dataset/iris.data")
```

Any command-line input or output is written as follows:

```
mkdir H2O_POJO
cd H2O_POJO
```

Bold: Indicates a new term, an important word, or words that you see onscreen. For instance, words in menus or dialog boxes appear in **bold**. Here is an example: "You can simply click the **Download POJO** button to download the model as a POJO."

> **Tips or Important Notes**
> Appear like this.

Get in touch

Feedback from our readers is always welcome.

General feedback: If you have questions about any aspect of this book, email us at customercare@packtpub.com and mention the book title in the subject of your message.

Errata: Although we have taken every care to ensure the accuracy of our content, mistakes do happen. If you have found a mistake in this book, we would be grateful if you would report this to us. Please visit www.packtpub.com/support/errata and fill in the form.

Piracy: If you come across any illegal copies of our works in any form on the internet, we would be grateful if you would provide us with the location address or website name. Please contact us at copyright@packt.com with a link to the material.

If you are interested in becoming an author: If there is a topic that you have expertise in and you are interested in either writing or contributing to a book, please visit authors.packtpub.com.

Share Your Thoughts

Once you've read *Practical Automated Machine Learning Using H2O.ai*, we'd love to hear your thoughts! Scan the QR code below to go straight to the Amazon review page for this book and share your feedback.

https://packt.link/r/1-801-07452-6

Your review is important to us and the tech community and will help us make sure we're delivering excellent quality content.

Part 1
H2O AutoML Basics

The objective of this part is to help you implement an easy, bare-bones demo of how to install, set up, and use H2O AutoML, opening up further exploration of and experimentation with the technology.

This section comprises the following chapters:

- *Chapter 1, Understanding H2O AutoML Basics*
- *Chapter 2, Working with H2O Flow (H2O's Web UI)*

Understanding H2O AutoML Basics

Machine Learning (**ML**) is the process of building analytical or statistical models using computer systems that learn from historical data and identify patterns in them. These systems then use these patterns and try to make predictive decisions that can provide value to businesses and research alike. However, the sophisticated mathematical knowledge required to implement an ML system that can provide any concrete value has discouraged several people from experimenting with it, leaving tons of undiscovered potential that they could have benefited from.

Automated Machine Learning (**AutoML**) is one of the latest ML technologies that has accelerated the adoption of ML by organizations of all sizes. It is the process of automating all these complex tasks involved in the ML life cycle. AutoML hides away all these complexities and automates them behind the scenes. This allows anyone to easily implement ML without any hassle and focus more on the results.

In this chapter, we will learn about one such AutoML technology by H2O.ai (`https://www.h2o.ai/`), which is simply named H2O AutoML. We will provide a brief history of AutoML in general and what problems it solves, as well as a bit about H2O.ai and its H2O AutoML technology. Then, we will code a simple ML implementation using H2O's AutoML technology and build our first ML model.

By the end of this chapter, you will understand what exactly AutoML is, the company H2O.ai, and its technology H2O AutoML. You will also understand what minimum requirements are needed to use H2O AutoML, as well as how easy it is to train your very first ML model using H2O AutoML without having to understand any complex mathematical rocket science.

In this chapter, we are going to cover the following topics:

- Understanding AutoML and H2O AutoML
- Minimum system requirements to use H2O AutoML
- Installing Java
- Basic implementation of H2O using Python
- Basic implementation of H2O using R
- Training your first ML model using H2O AutoML

Technical requirements

For this chapter, you will need the following:

- A decent web browser (Chrome, Firefox, or Edge), the latest version of your preferred web browser.
- An **Integrated Development Environment** (**IDE**) of your choice
- Jupyter Notebook by Project Jupyter (`https://jupyter.org/`) (optional)

All the code examples for this chapter can be found on GitHub at `https://github.com/PacktPublishing/Practical-Automated-Machine-Learning-on-H2O/tree/main/Chapter%201`.

Understanding AutoML and H2O AutoML

Before we begin our journey with H2O AutoML, it is important to understand what exactly AutoML is and what part it plays in the entire ML pipeline. In this section, we will try to understand the various steps involved in the ML pipeline and where AutoML fits into it. Then, we will explore what makes H2O's AutoML so unique among the various AutoML technologies.

Let's start by learning a bit about AutoML in general.

AutoML

AutoML is the process of automating the various steps that are performed while developing a viable ML system for predictions. A typical ML pipeline consists of the following steps:

1. **Data Collection**: This is the very first step in an ML pipeline. Data is collected from various sources. The sources can generate different types of data, such as categorical, numeric, textual, time series, or even visual and auditory data. All these types of data are aggregated together based on the requirements and are merged into a common structure. This could be a comma-separated value file, a parquet file, or even a table from a database.

2. **Data Exploration**: Once data has been collected, it is explored using basic analytical techniques to identify what it contains, the completeness and correctness of the data, and if the data shows potential patterns that can build a model.

3. **Data Preparation**: Missing values, duplicates, and noisy data can all affect the quality of the model as they introduce incorrect learning. Hence, the raw data that is collected and explored needs to be pre-processed to remove all anomalies using specific data processing methods.

4. **Data Transformation**: A lot of ML models work with different types of data. Some can work with categorical data, while some can only work with numeric data. That is why you may need to convert certain types of data from one form into the other. This allows the dataset to be fed properly during model training.

5. **Model Selection**: Once the dataset is ready, an ML model is selected to be trained. The model is chosen based on what type of data the dataset contains, what information is to be extracted from the dataset, as well as which model suits the data.

6. **Model Training**: This is where the model is trained. The ML system will learn from the processed dataset and create a model. This training can be influenced by several factors, such as data attribute weighting, learning rate, and other hyperparameters.

7. **Hyperparameter Tuning**: Apart from model training, another factor that needs to be considered is the model's architecture. The model's architecture depends on the type of algorithm used, such as the number of trees in a random forest or neurons in a neural network. We don't immediately know which architecture is optimal for a given model, so experimentation is needed. The parameters that define the architecture of a model are called hyperparameters; finding the best combination of hyperparameter values is known as **hyperparameter tuning**.

8. **Prediction**: The final step of the ML pipeline is prediction. Based on the patterns in the dataset that were learned by the model during training, the model can now make a generalized prediction on unseen data.

For non-experts, all these steps and their complexities can be overwhelming. Every step in the ML pipeline process has been developed over years of research and there are vast topics within themselves. AutoML is the process that automates the majority of these steps, from data exploration to hyperparameter tuning, and provides the best possible models to make predictions on. This helps companies focus on solving real-world problems with results rather than ML processes and workflows.

Now that you understand the different steps in an ML pipeline and how the steps are automated by AutoML, let's see why H2O's AutoML technology is one of the leading technologies in the industry.

H2O AutoML

H2O AutoML is an AutoML software technology developed by H2O.ai that simplifies how ML systems are developed by providing user-friendly interfaces that help non-experts experiment with ML. It is an in-memory, distributed, fast, and scalable ML and analytics platform that works on big data and can be used for enterprise needs.

It is written in Java and uses key-value storage to access data, models, and other ML objects that are involved. It runs on a cluster system and uses the multi-threaded MapReduce framework to parallelize data operations. It is also easy to communicate with it as it uses simple REST APIs. Finally, it has a web interface that provides a detailed graphical view of data and model details.

Not only does H2O AutoML automate the majority of the sophisticated steps involved in the ML life cycle, but it also provides a lot of flexibility for even expert data scientists to implement specialized model training processes. H2O AutoML provides a simple wrapper function that encapsulates several of the model training tasks that would otherwise be complicated to orchestrate. It also has extensive explainability functions that can describe the various details of the model training life cycle. This provides easy-to-export details of the models that users can use to explain the performance and justifications of the models that have been trained.

The best part about H2O AutoML is that it is entirely open source. You can find H2O's source code at `https://github.com/h2oai`. It is actively maintained by a community of developers serving in both open as well as closed sources companies. At the time of writing, it is on its third major version, which indicates that it is quite a mature technology and is feature-intensive – that is, it supports several major companies in the world. It also supports several programming languages, including R, Scala, Python, and Java, that can run on several operating systems and provides support for a wide variety of data sources that are involved in the ML life cycle, such as Hadoop Distributed File System, Hive, Amazon S3, and even **Java Database Connectivity (JDBC)**.

Now that you understand the basics of AutoML and how powerful H2O AutoML is, let's see what the minimum requirements are for a system to run H2O AutoML without any performance issues.

Minimum system requirements to use H2O AutoML

H2O is very easy to install, but certain minimum standard requirements need to be met for it to run smoothly and efficiently. The following are some of the minimum requirements needed by H2O in terms of hardware capabilities, along with other software support:

- The minimum hardware required by H2O is as follows:

  - **Memory**: H2O runs on an in-memory architecture, so it is limited by the physical memory of the system that uses it. Thus, to be able to process huge chunks of data, the more memory the system, has the better.

  - **Central Processing Unit (CPU)**: By default, H2O will use the maximum available CPUs of the system. However, at a minimum, it will need 4 CPUs.

  - **Graphical Processing Unit (GPU)**: GPU support is only available for XGBoost models in AutoML if the GPUs are NVIDIA GPUs (GPU Cloud, DGX Station, DGX-1, or DGX-2) or if it is a CUDA 8 GPU.

- The operating systems that support H2O are as follows:

  - **Ubuntu 12.04**

  - **OS X 10.9 or later**

  - **Windows 7 or later**

  - **CentOS 6 or later**

- The programming languages that support H2O are as follows:

  - **Java**: Java is mandatory for H2O. H2O requires a 64-bit JDK to build H2O and a 64-bit JRE to run its binary:

    - **Java versions supported**: Java SE 15, 14, 13, 12, 11, 10, 9, and 8

  - **Other Languages**: The following languages are only required if H2O is being run in those environments:

    - Python 2.7.x, 3.5.x, or 3.6.x

- Scala 2.10 or later

- R version 3 or later

- **Additional requirements**: The following requirements are only needed if H2O is being run in these environments:

  - **Hadoop**: Cloudera CDH 5.4 or later, Hortonworks HDP 2.2 or later, MapR 4.0 or later, or IBM Open Platform 4.2

  - **Conda**: 2.7, 3.5, or 3.6

  - **Spark**: Version 2.1, 2.2, or 2.3

Once we have a system that meets the minimum requirements, we need to focus on H2O's functional dependencies on other software. H2O has only one dependency and that is Java. Let's see why Java is important for H2O and how we can download and install the correct supported Java version.

Installing Java

H2O's core code is written in Java. It needs **Java Runtime Environment** (**JRE**) installed in your system to spin up an H2O server cluster. H2O also trains all the ML algorithms in a multi-threaded manner, which uses the Java Fork/Join framework on top of its MapReduce framework. Hence, having the latest Java version that is compatible with H2O to run H2O smoothly is highly recommended.

You can install the latest stable version of Java from `https://www.oracle.com/java/technologies/downloads/`.

When installing Java, it is important to be aware of which bit version your system runs on. If it is a 64-bit version, then make sure you are installing the 64-bit Java version for your operating system. If it is 32-bit, then go for the 32-bit version.

Now that we have installed the correct Java version, we can download and install H2O. Let's look at a simple example of how we can do that using Python.

Basic implementation of H2O using Python

Python is one of the most popular languages in the ML field of computer programming. It is widely used in all industries and has tons of actively maintained ML libraries that provide a lot of support in creating ML pipelines.

We will start by installing the Python programming language and then installing H2O using Python.

Installing Python

Installing Python is very straightforward. It does not matter whether it is Python 2.7 or Python 3 and above as H2O works completely fine with both versions of the language. However, if you are using anything older than Python 2.7, then you will need to upgrade your version.

It is best to go with Python 3 as it is the current standard and Python 2.7 is outdated. Along with Python, you will also need `pip`, Python's package manager. Now, let's learn how to install Python on various operating systems:

- On Linux (Ubuntu, Mint, Debian):

    - For Python 2.7, run the following command in the system Terminal:

        ```
        sudo apt-get python-pip
        ```

    - For Python 3, run the following command in the system Terminal:

        ```
        sudo apt-get python3-pip
        ```

- On macOS: macOS version 10.8 comes with Python 2.7 pre-installed. If you want to install Python 3, then go to https://python.org, go to the **Downloads** section, and download the latest version of Python 3 for macOS.

- On Windows: Unlike macOS, Windows does not come with any pre-installed Python language support. You will need to download a Python installer for Windows from https://python.org. The installer will depend on your Windows operating system – that is, if it is 64-bit or 32-bit.

Now that you know how to install the correct version of Python, let's download and install the H2O Python module using Python.

Installing H2O using Python

H2O has a Python module available in the Python package index. To install the h2o Python module, all you need to do is to execute the following command in your Terminal:

```
pip install h2o
```

And that's pretty much it.

To test if it has been successfully downloaded and installed, follow these steps:

1. Open your Python Terminal.

2. Import the h2o module by running the following command:

    ```
    import h2o
    ```

3. Initialize H2O to spin up a local h2o server by running the following command:

    ```
    h2o.init()
    ```

The following screenshot shows the results you should get after initializing h2o:

```
Checking whether there is an H2O instance running at http://localhost:54321 ..... not found.
Attempting to start a local H2O server...
; Java HotSpot(TM) 64-Bit Server VM (build 17.0.2+8-LTS-86, mixed mode, sharing)
  Starting server from C:\Users\ajgao\AppData\Local\Programs\Python\Python310\Lib\site-packages\h2o\backend\bin\h2o.jar
  Ice root: C:\Users\ajgao\AppData\Local\Temp\tmpmswmzso9
  JVM stdout: C:\Users\ajgao\AppData\Local\Temp\tmpmswmzso9\h2o_ajgao_started_from_python.out
  JVM stderr: C:\Users\ajgao\AppData\Local\Temp\tmpmswmzso9\h2o_ajgao_started_from_python.err
  Server is running at http://127.0.0.1:54321
Connecting to H2O server at http://127.0.0.1:54321 ... successful.

H2O_cluster_uptime:          01 secs
H2O_cluster_timezone:        Europe/Dublin
H2O_data_parsing_timezone:   UTC
H2O_cluster_version:         3.36.1.2
H2O_cluster_version_age:     6 days
H2O_cluster_name:            H2O_from_python_ajgao_etoqk4
H2O_cluster_total_nodes:     1
H2O_cluster_free_memory:     3.963 Gb
H2O_cluster_total_cores:     12
H2O_cluster_allowed_cores:   12
H2O_cluster_status:          locked, healthy
H2O_connection_url:          http://127.0.0.1:54321
H2O_connection_proxy:        {"http": null, "https": null}
H2O_internal_security:       False
Python_version:              3.10.2 final

>>>
```

Figure 1.1 – H2O execution using Python

Let's have a quick look at the output we got. First, it ran successfully, so mission accomplished.

After executing h2o.init() by reading the output logs, you will see that H2O checked if there is already an H2O server instance running on localhost with port 54321. In this scenario, there wasn't any H2O server instance running previously, so H2O attempted to start a local server on the same port. If it had found an already existing local H2O instance on the port, then it would have reused the same instance for any further H2O command executions.

Then, it used Java version 16 to start the H2O instance. You may see a different Java version, depending on which version you have installed in your system.

Next, you will see the location of the h2o jar file that the server was started from, followed by the location of the **Java Virtual Machine (JVM)** logs.

Once the server is up and running, it shows the URL of the H2O server locally hosted on your system and the status of the H2O Python library's connection to the server.

Lastly, you will see some basic metadata regarding the server's configuration. This metadata may be slightly different from what you see in your execution as it depends a lot on the specifications of your system. For example, by default, H2O will use all the cores available on your system for processing. So, if you have an 8-core system, then the `H2O_cluster_allowed_cores` property value will be 8. Alternatively, if you decide to use only four cores, then you can execute `h2o.init(nthreads=4)` to use only four cores, reflecting the same in the server configuration output.

Now that you know how to implement H2O using Python, let's learn how to do the same in the R programming language.

Basic implementation of H2O using R

The R programming language is a very popular language in the field of ML and data science because of its extensive support for statistical and data manipulation operations. It is widely used by data scientists and data miners for developing analytical software.

We will start by installing the R programming language and then installing H2O using R.

Installing R

An international team of developers maintains the R programming language. They have a dedicated web page for the R programming language called **The Comprehensive R Archive Network (CRAN)**: `https://cran.r-project.org/`. There are different ways of installing R, depending on what operating system you use:

- On Linux (Ubuntu, Mint, Debian):

  Execute the following command in the system Terminal:

  ```
  sudo apt-get install r-base
  ```

- On macOS: To install R, go to `https://cran.r-project.org/`, go to the **Download R for macOS** hyperlink, and download the latest release of R for macOS.

- On Windows: Similar to how you install R on macOS, you can download the `.exe` file from `https://cran.r-project.org/`, go to the **Download R for Windows** hyperlink, and download the latest release of R for Windows.

Another great way of installing R on macOS and Windows is through RStudio. RStudio simplifies the installation of R-supported software and is also a very good IDE for R programming in general. You can download R studio from https://www.rstudio.com/.

Now that you know how to install the correct version of R, let's download and install the H2O R package using the R programming language.

Installing H2O using R

Similar to Python, H2O provide support for the R programming language as well.

To install the R packages, follow these steps:

1. First, we need to download the H2O R package dependencies. For this, execute the following command in your R Terminal:

    ```
    install.packages(c("RCurl", "jsonlite"))
    ```

2. Then, to install the actual h2o package, execute the following command in your R Terminal:

    ```
    install.packages("h2o")
    ```

    And you are done.

3. To test if it has been successfully downloaded and installed, open your R Terminal, import the h2o library, and execute the h2o.init() command. This will spin up a local H2O server.

The results can be seen in the following screenshot:

```
H2O is not running yet, starting it now...

Note:  In case of errors look at the following log files:
    C:\Users\ajgao\AppData\Local\Temp\RtmpGOgnlL\filee50164e6416/h2o_ajgao_started_from_r.out
    C:\Users\ajgao\AppData\Local\Temp\RtmpGOgnlL\filee50756a39d3/h2o_ajgao_started_from_r.err

java version "17.0.2" 2022-01-18 LTS
Java(TM) SE Runtime Environment (build 17.0.2+8-LTS-86)
Java HotSpot(TM) 64-Bit Server VM (build 17.0.2+8-LTS-86, mixed mode, sharing)

Starting H2O JVM and connecting:  Connection successful!

R is connected to the H2O cluster:
    H2O cluster uptime:         1 seconds 849 milliseconds
    H2O cluster timezone:       Europe/Dublin
    H2O data parsing timezone:  UTC
    H2O cluster version:        3.36.1.2
    H2O cluster version age:    6 days
    H2O cluster name:           H2O_started_from_R_ajgao_kcp486
    H2O cluster total nodes:    1
    H2O cluster total memory:   3.96 GB
    H2O cluster total cores:    12
    H2O cluster allowed cores:  12
    H2O cluster healthy:        TRUE
    H2O Connection ip:          localhost
    H2O Connection port:        54321
    H2O Connection proxy:       NA
    H2O Internal Security:      FALSE
    R Version:                  R version 3.6.1 (2019-07-05)

>
```

Figure 1.2 – H2O execution using R

Let's have a quick look at the output we got.

After executing h2o.init(), the H2O client will check if there is an H2O server instance already running on the system. The H2O server is usually run locally on port 54321 by default. If it had found an already existing local H2O instance on the port, then it would have reused the same instance. However, in this scenario, there wasn't any H2O server instance running on port 54321, which is why H2O attempted to start a local server on the same port.

Next, you will see the location of the JVM logs. Once the server is up and running, the H2O client tries to connect to it and the status of the connection to the server is displayed. Lastly, you will see some basic metadata regarding the server's configuration. This metadata may be slightly different from what you see in your execution as it depends a lot on the specifications of your system. For example, by default, H2O will use all the cores available on your system for processing. So, if you have an 8-core system, then the H2O_cluster_allowed_cores property value will be 8. Alternatively, if you decide to use only four cores, then you can execute the h2o.init(nthreads=4) command to use only four cores, thus reflecting the same in the server configuration output.

Now that you know how to implement H2O using Python and R, let's create our very first ML model and make predictions on it using H2O AutoML.

Training your first ML model using H2O AutoML

All ML pipelines, whether they're automated or not, eventually follow the same steps that were discussed in the *Understanding AutoML and H2O AutoML* section in this chapter.

For this implementation, we will be using the Iris flower dataset. This dataset can be found at https://archive.ics.uci.edu/ml/datasets/iris.

Understanding the Iris flower dataset

The Iris flower dataset, also known as Fisher's Iris dataset, is one of the most popular multivariate datasets – that is, a dataset in which there are two or more variables that are analyzed per observation during model training. The dataset consists of 50 samples of three different varieties of the Iris flower. The features in the dataset include the length and width of the petals and sepals in centimeters. The dataset is often used for studying various classification techniques in ML because of its simplicity. The classification is performed by using the length and width of the petals and sepals as features that determine the class of the Iris flower.

The following screenshot shows a small sample of the dataset:

C1	C2	C3	C4	C5
5.1	3.5	1.4	0.2	Iris-setosa
4.9	3	1.4	0.2	Iris-setosa
4.7	3.2	1.3	0.2	Iris-setosa
4.6	3.1	1.5	0.2	Iris-setosa
5	3.6	1.4	0.2	Iris-setosa
5.4	3.9	1.7	0.4	Iris-setosa
4.6	3.4	1.4	0.3	Iris-setosa
5	3.4	1.5	0.2	Iris-setosa
4.4	2.9	1.4	0.2	Iris-setosa
4.9	3.1	1.5	0.1	Iris-setosa

Figure 1.3 – Iris dataset

The columns in the dataset represent the following:

- **C1**: Sepal length in cm
- **C2**: Sepal width in cm
- **C3**: Petal length in cm
- **C4**: Petal width in cm
- **C5**: Class:

  - Iris-setosa
  - Iris-versicolour
  - Iris-virginica

In this scenario, **C1**, **C2**, **C3**, and **C4** represent the features that are used to determine **C5**, the class of the Iris flower.

Now that you understand the contents of the dataset that we are going to be working with, let's implement our model training code.

Model training

Model training is the process of finding the best combination of biases and weights for a given ML algorithm so that it minimizes a loss function. A **loss function** is a way of measuring how far the predicted value is from the actual value. So, minimizing it indicates that the model is getting closer to making accurate predictions – in other words, it's learning. The ML algorithm builds a mathematical representation of the relationship between the various features in the dataset and the target label. Then, we use this mathematical representation to predict the potential value of the target label for certain feature values. The accuracy of the predicted values depends a lot on the quality of the dataset, as well as the combination of weights and biases against features used during model training. However, all of this is entirely automated by AutoML and, as such, is not a concern for us.

With that in mind, let's learn how to quickly and easily create an ML model using H2O in Python.

Model training and prediction in Python

The H2O Python module makes it easy to use H2O in a Python program. The inbuilt functions in the H2O Python module are straightforward to use and hide away a lot of the complexities of using H2O.

Follow these steps to train your very first model in Python using H2O AutoML:

1. Import the H2O module:

    ```
    import h2o
    ```

2. Initialize H2O to spin up a local H2O server:

    ```
    h2o.init()
    ```

    The `h2o.init()` command starts up or reuses an H2O server instance running locally on port 54321.

3. Now, you can import the dataset by using the `h2o.import_file()` command while passing the location of the dataset into your system.

4. Next, import the dataset by passing the location where you downloaded the dataset:

    ```
    data = h2o.import_file("Dataset/iris.data")
    ```

5. Now, you need to identify which columns of the DataFrame are the features and which are the labels. A **label** is something that we want to predict, while **features** are attributes of the label that help identify the label. We train models on these features and then predict the value of the label, given a specific set of feature values. Referring to the dataset in the *Understanding the Iris flower dataset* section, let's set all the column names – C1, C2, C3, C4, and C5 – as a list of features:

    ```
    features = data.columns
    ```

6. Based on our DataFrame, the C5 column, which denotes the class of the Iris flower, is the column that we want to eventually predict once the model has been trained. Hence, we denote C5 as the label and remove it from the remaining set of column names, which we will note as features. Set the target label and remove it from the list of features:

    ```
    label = "C5"
    features.remove(label)
    ```

7. Split the DataFrame into training and testing DataFrames:

    ```
    train_dataframe, test_dataframe = data.split_frame([0.8])
    ```

    The data.split_frame([0.8]) command splits the DataFrame into two – a training DataFrame and another for testing. The training DataFrame contains 80% of the data, while the testing DataFrame contains the remaining 20%. We will use the training DataFrame to train the model and the testing DataFrame to run predictions on the model once it has been trained to test how the model performs.

    > **Tip**
    >
    > If you are curious as to how H2O splits the dataset based on ratios and how it randomizes the data between different splits, feel free to explore and experiment with the split_frame function. You can find more details at https://docs.h2o.ai/h2o/latest-stable/h2o-py/docs/_modules/h2o/frame.html#H2OFrame.split_frame.

8. Initialize the H2O AutoML object. Here, we have set the max_model parameter to 10 to limit the number of models that will be trained by H2O, set AutoML to 10, and set the random seed generator to 1:

    ```
    aml=h2o.automl.H2OAutoML(max_models=10, seed = 1)
    ```

9. Now, trigger the AutoML training by passing in the feature columns – that is C1, C2, C3, and C4 – in (*x*), the label column C5 in (*y*), and the train_dataframe DataFrame using the aml.train() command. This is when H2O starts its automated model training.

10. Train the model using the H2O AutoML object:

```
aml.train(x = features, y = label, training_frame =
train_dataframe)
```

During the training, H2O will analyze the type of the label column. For numerical labels, H2O treats the ML problem as a regression problem. If the label is categorical, then it treats the problem as a classification problem. For the Iris flower dataset, the C5 column is a categorical value containing class values. H2O will analyze this column and correctly identify that it is a classification problem and train classification models.

H2O AutoML trains several models behind the scenes using different types of ML algorithms. All the models that have been trained are evaluated on the test dataset and their performance is measured. H2O also provides detailed information about all the models, which users can use to further experiment on the data or compare different ML algorithms and understand which ones are more suitable to solve their ML problem. H2O can end up training 20-30 models, which can take a while. However, since we have passed the max_models parameter as 10, we are limiting the number of models that will be trained so that we can see the output of the training process quickly. More on ensemble learning will be discussed in *Chapter 5, Understanding AutoML Algorithms*.

11. Once the training has finished, AutoML creates a Leaderboard of all the models it has trained, ranking them from the best performing to the worst. This ranking is achieved by comparing all the models' error metrics. **Error metrics** are values that measure how many errors the model makes when making predictions on a sample test dataset with the actual label values. Lower error metrics indicate that the model makes fewer errors during prediction, which indicates that it is a better model compared to one with a higher error metric. Extract the AutoML Leaderboard:

```
model_leaderboard = aml.leaderboard
```

12. Display the AutoML Leaderboard:

```
model_leaderboard.head(rows=model_leaderboard.nrows)
```

The Leaderboard will look as follows:

model_id	mean_per_class_error	logloss	rmse	mse
GLM_1_AutoML_1_20211221_224844	0.0254274	0.0730056	0.148617	0.0220871
StackedEnsemble_BestOfFamily_1_AutoML_1_20211221_224844	0.0254274	0.0889165	0.155702	0.0242432
StackedEnsemble_BestOfFamily_3_AutoML_1_20211221_224844	0.034188	0.234761	0.244628	0.0598429
StackedEnsemble_AllModels_4_AutoML_1_20211221_224844	0.0418803	0.212898	0.207104	0.042892
GBM_5_AutoML_1_20211221_224844	0.0423077	0.153568	0.196082	0.038448
XGBoost_3_AutoML_1_20211221_224844	0.0423077	0.171005	0.205224	0.0421171
StackedEnsemble_BestOfFamily_5_AutoML_1_20211221_224844	0.0423077	0.272447	0.208633	0.0435279
StackedEnsemble_AllModels_1_AutoML_1_20211221_224844	0.0425214	0.239748	0.247898	0.0614534
StackedEnsemble_BestOfFamily_2_AutoML_1_20211221_224844	0.0425214	0.242794	0.250567	0.0627836
GBM_2_AutoML_1_20211221_224844	0.0508547	0.164525	0.20544	0.0422055
XRT_1_AutoML_1_20211221_224844	0.0508547	0.154602	0.196747	0.0387094
DRF_1_AutoML_1_20211221_224844	0.0508547	0.155568	0.201598	0.0406418
XGBoost_2_AutoML_1_20211221_224844	0.0508547	0.230482	0.231617	0.0536463
GBM_4_AutoML_1_20211221_224844	0.0508547	0.159493	0.201367	0.0405486
StackedEnsemble_BestOfFamily_6_AutoML_1_20211221_224844	0.0508547	0.140577	0.198154	0.0392648
GBM_3_AutoML_1_20211221_224844	0.0508547	0.158446	0.203925	0.0415853
StackedEnsemble_AllModels_5_AutoML_1_20211221_224844	0.0508547	0.13417	0.197749	0.0391048
StackedEnsemble_AllModels_2_AutoML_1_20211221_224844	0.0515304	0.244011	0.252697	0.0638558
StackedEnsemble_AllModels_3_AutoML_1_20211221_224844	0.059188	0.183036	0.221313	0.0489796
StackedEnsemble_BestOfFamily_4_AutoML_1_20211221_224844	0.059188	0.246594	0.239717	0.0574643
XGBoost_1_AutoML_1_20211221_224844	0.0925214	0.478273	0.385523	0.148628

Figure 1.4 – H2O AutoML Leaderboard (Python)

The Leaderboard includes the following details:

- `model_id`: This represents the ID of the model.
- `mean_per_class_error`: This metric is used to measure the average of the errors of each class in your multi-class dataset.
- `logloss`: This metric is used to measure the negative average of the log of corrected predicted probabilities for each instance.
- **Root Mean Squared Error (RMSE)**: This metric is used to measure the standard deviation of prediction errors.
- **Mean Squared Error (MSE)**: This metric is used to measure the average of the squares of the errors.

The Leaderboard sorts the models based on certain default metrics, depending on the type of ML problem, unless specifically mentioned during AutoML training. The Leaderboard sorts the models based on the **AUC** metric for binary classification, `mean_per_class_error` for multinomial classification, and **deviance** for regression.

The metrics are different measures of error in the model's performance. So, the smaller the error value, the better the model is for making accurate predictions. We will explore the different model performance metrics in *Chapter 6, Understanding H2O AutoML Leaderboard and Other Performance Metrics*.

In this case, `GLM_1_AutoML_1_20211221_224844` is the best model according to H2O AutoML since it is a multinomial classification problem and this model has the lowest `mean_per_class_error`.

You may notice that despite passing the `max_model` value as `10`, when triggering AutoML for training, we see more than 10 models in the Leaderboard. This is because only 10 models have been trained; the remaining models are Stacked Ensemble models. **Stacked Ensemble** models are models that are created from what other models have learned and are not technically trained in the normal sense. We will learn more about Stacked Ensemble models in *Chapter 5, Understanding AutoML Algorithms*, and more about the Leaderboard in *Chapter 6, Understanding H2O AutoML Leaderboard and Other Performance Metrics*.

Congratulations! You have officially trained your very first ML model using H2O AutoML and it is now ready to be used to make predictions.

Making predictions is very straightforward: we will use the `test_dataframe` DataFrame that was created from the `data.split_frame([0.8])` command.

Execute the following command in Python:

```
aml.predict(test_dataframe)
```

That's it – everything is wrapped inside the `predict` function of the model object.

After executing the prediction, you will see the following results:

predict	Iris-setosa	Iris-versicolor	Iris-virginica
Iris-setosa	0.996763	0.0029518	0.000284888
Iris-setosa	0.999722	0.000171882	0.000106306
Iris-setosa	0.99952	0.000345017	0.0001354
Iris-setosa	0.999739	0.000157519	0.000103133
Iris-setosa	0.999975	2.05119e-06	2.28322e-05
Iris-setosa	0.999801	0.000108085	9.04886e-05
Iris-setosa	0.999452	0.000405189	0.00014317
Iris-setosa	0.999515	0.000349113	0.000135955
Iris-setosa	0.999335	0.000509734	0.000155041
Iris-setosa	0.999627	0.000251481	0.000121322

Figure 1.5 – H2O AutoML model prediction (Python)

The prediction result shows a table where every row is a representation of predictions for the rows present in the test DataFrame. The `predict` column indicates what Iris class it is for that row, while the remaining columns are the calculated probabilities of the Iris classes, as denoted in the column's name, by the model after reading the feature values of that row. In short, the model predicts that for *row 1*, there is a *99.6763%* chance that it is Iris-setosa.

Congratulations! You have now made an accurate prediction using your newly trained model using AutoML.

Now that we've seen how easy it is to use H2O AutoML in Python, let's learn how to do the same in the R programming language.

Model training and prediction in R

Similar to Python, training and making predictions using H2O AutoML in the R programming language is also very easy. H2O has a lot of support for the R programming language and, as such, has encapsulated much of the sophistication of ML behind ready-to-use functions.

Let's look at a model training example that uses H2O AutoML in the R programming language on the Iris flower dataset.

You will notice that training models in R is similar to how we do it in Python, with the only difference being the slight change in syntax.

Follow these steps:

1. Import the H2O library:

    ```
    library(h2o)
    ```

2. Initialize H2O to spin up a local H2O server:

    ```
    h2o.init()
    ```

    `h2o.init()` will start up an H2O server instance that's running locally on port 54321 and connect to it. If an H2O server already exists on the same port, then it will reuse it.

3. Import the dataset using `h2o.importFile("Dataset/iris.data")` while passing the location of the dataset in your system as a parameter. Import the dataset:

    ```
    data <- h2o.importFile("Dataset/iris.data")
    ```

4. Now, you need to set which columns of the dataframe are the features and which column is the label. Set the C5 column as the target label and the remaining column names as the list of features:

    ```
    label <- "C5"
    features <- setdiff(names(data), label)
    ```

5. Split the DataFrame into two parts:

    ```
    parts <- h2o.splitFrame(data, 0.8)
    ```

    One DataFrame will be used for training, while the other will be used for testing/validating the model being trained. `parts <- h2o.splitFrame(data, 0.8)` splits the DataFrame into two parts. One DataFrame contains 80% of the data, while the other contains the remaining 20%. Now, assign the DataFrame that contains 80% of the data as the training DataFrame and the other as the testing or validation DataFrame.

6. Assign the first part as the training DataFrame:

```
train_dataframe <- parts[[1]]
```

7. Assign the second part as the testing DataFrame:

```
test_dataframe <- parts[[2]]
```

8. Now that the dataset has been imported and its features and labels have been identified, let's pass them to H2O's AutoML to train models. This means that you can implement the AutoML model training function in R using h2o.automl(). Train the model using H2O AutoML:

```
aml <- h2o.automl(x = features, y = label, training_frame
= train_dataframe, max_models=10, seed = 1)
```

9. Extract the AutoML Leaderboard:

```
model_leaderboard <- aml@leaderboard
```

10. Display the AutoML Leaderboard:

```
print(model_leaderboard, n = nrow(model_leaderboard))
```

Once the training has finished, AutoML will create a Leaderboard of all the models it has trained, ranking them from the best performing to the worst.

The Leaderboard will display the results as follows:

model_id	mean_per_class_error	logloss	rmse	mse
GBM_3_AutoML_8_20211222_02555	0.02503053	0.1441918	0.1790187	0.03204771
GLM_1_AutoML_8_20211222_02555	0.02503053	0.06051945	0.1367546	0.01870183
StackedEnsemble_AllModels_5_AutoML_8_20211222_02555	0.02503053	0.08676523	0.1584221	0.02509755
GBM_4_AutoML_8_20211222_02555	0.03296703	0.15864028	0.1871248	0.03501568
StackedEnsemble_BestOfFamily_2_AutoML_8_20211222_02555	0.03455433	0.22094574	0.2311208	0.05341682
StackedEnsemble_BestOfFamily_3_AutoML_8_20211222_02555	0.04090354	0.21032039	0.2249462	0.05060078
StackedEnsemble_BestOfFamily_4_AutoML_8_20211222_02555	0.04151404	0.15865856	0.1955133	0.03822546
StackedEnsemble_AllModels_4_AutoML_8_20211222_02555	0.04151404	0.18315141	0.1858142	0.03452691
StackedEnsemble_AllModels_3_AutoML_8_20211222_02555	0.04151404	0.17914264	0.2003599	0.0401441
StackedEnsemble_AllModels_1_AutoML_8_20211222_02555	0.04310134	0.2171082	0.2293443	0.05259883
StackedEnsemble_BestOfFamily_6_AutoML_8_20211222_02555	0.04310134	0.09428996	0.1672787	0.02798217
XRT_1_AutoML_8_20211222_02555	0.04884005	0.12547862	0.1850836	0.03425593
GBM_2_AutoML_8_20211222_02555	0.04884005	0.16726969	0.1934784	0.03743388
XGBoost_3_AutoML_8_20211222_02555	0.04884005	0.16601239	0.1983851	0.03935665
XGBoost_2_AutoML_8_20211222_02555	0.04884005	0.22718554	0.2309562	0.05334075
StackedEnsemble_BestOfFamily_5_AutoML_8_20211222_02555	0.04945055	0.26499193	0.221083	0.04887771
StackedEnsemble_BestOfFamily_1_AutoML_8_20211222_02555	0.05103785	0.10792744	0.187905	0.03530827
StackedEnsemble_AllModels_2_AutoML_8_20211222_02555	0.05103785	0.21228471	0.2263655	0.05124134
GBM_5_AutoML_8_20211222_02555	0.05738706	0.13549309	0.1926783	0.03712494
DRF_1_AutoML_8_20211222_02555	0.05738706	0.1279311	0.1848138	0.03415613
XGBoost_1_AutoML_8_20211222_02555	0.10989011	0.46102681	0.3760385	0.14140496

Figure 1.6 – H2O AutoML Leaderboard (R)

The Leaderboard includes the same details as we saw in the Leaderboard we got when training models in Python.

However, you may notice that the best model that's suggested in this Leaderboard is different from the one we got in our previous experiment.

In this case, GBM_3_AutoML_8_20211222_02555 is the best model according to H2O AutoML, while in the previous experiment, it was GLM_1_AutoML_1_20211221_224844. This may be due to several factors, such as a different random number being generated for the seed value during model training or different data values being split across the training and testing DataFrames between the two experiments. This is what makes ML tricky – every step that you perform in a model training pipeline can greatly affect the overall performance of your trained model. At the end of the day, ML is a best-effort approach to making the most accurate predictions.

Congratulations – you have officially trained your ML model using H2O AutoML in R. Now, let's learn how to make predictions on it. We will use the testing DataFrame we created after the split function to make predictions on the model we trained.

Execute the following command in R to make predictions:

```
predictions <- h2o.predict(aml, test_dataframe)
```

The predict function of the h2o object accepts two parameters. One is the model object, which in our case is the aml object, while the other is the DataFrame to make predictions on. By default, the aml object will use the best model in the Leaderboard to make predictions.

After executing the prediction, you will see the following results:

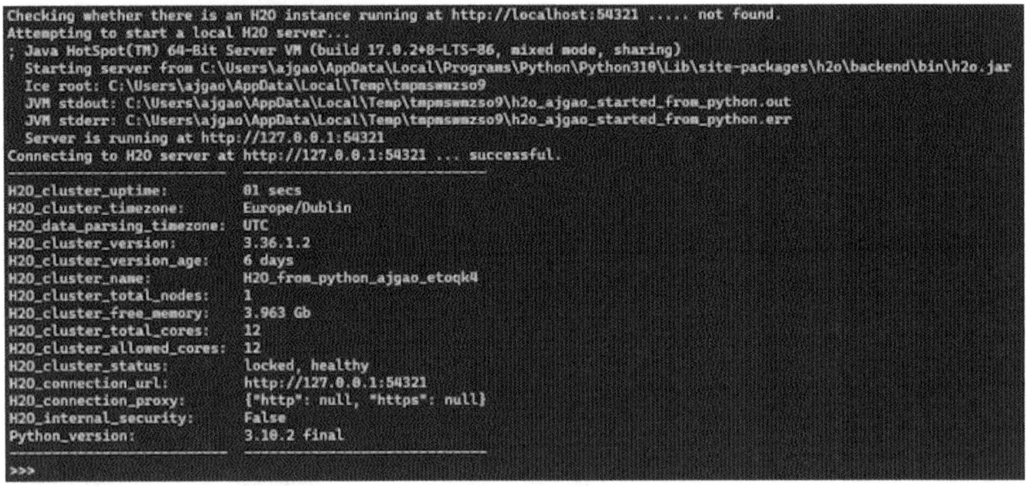

```
Checking whether there is an H2O instance running at http://localhost:54321 ..... not found.
Attempting to start a local H2O server...
; Java HotSpot(TM) 64-Bit Server VM (build 17.0.2+8-LTS-86, mixed mode, sharing)
  Starting server from C:\Users\ajgao\AppData\Local\Programs\Python\Python310\Lib\site-packages\h2o\backend\bin\h2o.jar
  Ice root: C:\Users\ajgao\AppData\Local\Temp\tmpmswmzso9
  JVM stdout: C:\Users\ajgao\AppData\Local\Temp\tmpmswmzso9\h2o_ajgao_started_from_python.out
  JVM stderr: C:\Users\ajgao\AppData\Local\Temp\tmpmswmzso9\h2o_ajgao_started_from_python.err
  Server is running at http://127.0.0.1:54321
Connecting to H2O server at http://127.0.0.1:54321 ... successful.

H2O_cluster_uptime:         01 secs
H2O_cluster_timezone:       Europe/Dublin
H2O_data_parsing_timezone:  UTC
H2O_cluster_version:        3.36.1.2
H2O_cluster_version_age:    6 days
H2O_cluster_name:           H2O_from_python_ajgao_etoqk4
H2O_cluster_total_nodes:    1
H2O_cluster_free_memory:    3.963 Gb
H2O_cluster_total_cores:    12
H2O_cluster_allowed_cores:  12
H2O_cluster_status:         locked, healthy
H2O_connection_url:         http://127.0.0.1:54321
H2O_connection_proxy:       {"http": null, "https": null}
H2O_internal_security:      False
Python_version:             3.10.2 final

>>>
```

Figure 1.7 – H2O AutoML model prediction (R)

The results show a table with similar details that we saw in our previous experiment with Python. Every row is a representation of predictions for the rows present in the test DataFrame. The `predict` column indicates what Iris class it is for that row, while the remaining columns are the calculated probabilities of the Iris classes.

Congratulations – you have made an accurate prediction using your newly trained model using AutoML in R. Now, let's summarize this chapter.

Summary

In this chapter, we understood the various steps in an ML pipeline and how AutoML plays a part in automating some of those steps. Then, we prepared our system to use H2O AutoML by installing the basic requirements. Once our system was ready, we implemented a simple application in Python and R that uses H2O AutoML to train a model on the Iris flower dataset. Finally, we understood the Leaderboard results and made successful predictions on the ML model that we just trained. All of this helped us test the waters of H2O AutoML, thus opening doors to more advanced concepts of H2O AutoML.

In the next chapter, we will explore H2O's web **User Interface** (**UI**) so that we can understand and observe various ML details using an interactive visual interface.

2
Working with H2O Flow (H2O's Web UI)

Machine Learning (**ML**) is more than just code. It involves tons of observations from different perspectives. As powerful as actual coding is, a lot of information gets hidden away behind the Terminal on which you code. Humans have always understood pictures more easily than words. Similarly, as complex as ML is, it can be very easy and fun to implement with the help of interactive **User Interfaces** (**UIs**). Working with a colorful UI over the dull black and white pixelated Terminal is always a plus when learning about difficult topics.

H2O Flow is a web-based UI developed by the H2O.ai team. This interface works on the same backend that we learned about in *Chapter 1, Understanding H2O AutoML Basics*. It is simply a web UI wrapped over the main H2O library, which passes inputs and triggers functions on the backend server and reads the results by displaying them back to the user.

In this chapter, we will learn how to work with H2O Flow. We will perform all the typical steps of the ML pipeline, which we learned about in the *Understanding AutoML and H2O AutoML* section of *Chapter 1, Understanding H2O AutoML Basics*, from reading datasets to making predictions using the trained models. Also, we will explore a few metrics and model details to help us ease into more advanced topics later. This chapter is hands-on, and we will learn about the various parts of H2O Flow as we create our ML pipeline.

By the end of this chapter, you will be able to navigate and use the various features of H2O Flow. Additionally, you will be able to train your ML models and use them for predictions without needing to write a single line of code using H2O Flow.

In this chapter, we are going to cover the following topics:

- Understanding the basics of H2O Flow
- Working with data functions in H2O Flow
- Working with model training functions in H2O Flow
- Working with prediction functions in H2O Flow

Technical requirements

You will require the following:

- A decent web browser (Chrome, Firefox, or Edge), the latest version of your preferred web browser.

Understanding the basics of H2O Flow

H2O Flow is an open source web interface that helps users execute code, plot graphs, and display dataframes on a single page called a **Flow notebook** or just **Flow**.

Users of Jupyter notebooks will find H2O Flow very similar. You write your executable code in cells, and the output of the code is displayed below it when you execute the cell. Then, the cursor moves on to the next cell. The best thing about a Flow is that it can be easily saved, exported, and imported between various users. This helps a lot of data scientists share results among various stakeholders, as they just need to save the execution results and share the flow.

In the following sub-sections, we will gain an understanding of the basics of H2O Flow. Let's begin our journey with H2O Flow by, first, downloading it to our system.

Downloading and launching H2O Flow

In order to run H2O Flow, you will need to first download the H2O Flow **Java Archive (JAR)** file onto your system, and then run the JAR file once it has been downloaded.

You can download and launch H2O Flow using the following steps:

1. You can download H2O Flow at `https://h2o-release.s3.amazonaws.com/h2o/master/latest.html`.

2. Once the ZIP file has been downloaded, open a Terminal and run the following commands in the folder where you downloaded the ZIP file:

    ```
    unzip {name_of_the_h2o_zip_file}
    ```

3. To run H2O Flow, run the following command inside the folder of your recently unzipped h2o file:

    ```
    java -jar h2o.jar
    ```

This will start an H2O Web UI on `http://localhost:54321`.

Now that we have downloaded and launched H2O Flow, let's briefly explore it to get an understanding of what functionalities it has to offer.

Exploring H2O Flow

H2O Flow is a very feature-intensive application. It has almost all the features you will need to create your ML pipeline. Considering the various steps involved in an ML pipeline, H2O Flow provides plenty of functionality for all of these steps. This can be overwhelming for a lot of people. Therefore, it is worthwhile exploring the application and focusing on specific parts of the application one at a time.

In the following sections, we will learn about all of these parts in detail; however, some functions might be too complex to understand at this stage. We will understand them better in the upcoming chapters. For the time being, we will focus on getting an overview of how we can use H2O Flow to create our ML pipeline.

So, let's begin with the exploration by, first, launching H2O Flow and opening your web browser at the `http://localhost:54321` URL.

The following screenshot shows you the main page of the H2O Flow UI:

Figure 2.1 – The main page of the H2O Flow UI

Don't be alarmed if you see something slightly different. This might be because the version you installed might be different from the one shown in this book. Nevertheless, the basic setup of the UI should be similar. As you can see, there are plenty of interactive options to choose from. This can be a little overwhelming at first, so let's take it piece by piece.

At the very top of the web page, there are various ML object operations, each with its own drop-down list of functions. They are as follows:

- **Flow**: This section contains all operations related to the Flow notebook.

- **Cell**: This section contains all operations related to the individual cells in the notebook.

- **Data**: This section contains all operations related to data manipulations.

- **Model**: This section contains all operations related to model training and algorithms.

- **Score**: This section contains all operations related to scoring and making predictions.

- **Admin**: This section contains all administrative operations such as downloading logs.

- **Help**: This section contains all links to H2O documentation for additional details.

We will go through them in greater detail, step by step, once we start creating our ML pipeline later.

The following screenshot shows you the various ML function operations of the H2O Flow UI:

Figure 2.2 – The ML object operation toolbar

Following this, we have the **Flow Name** and **Flow Toolbar** settings. By default, the flow name will be **Untitled Flow**. The flow name helps identify the whole flow in general so that you do not mix the different experiments. The **Flow Toolbar** section is a simple toolbar that helps you with basic operations such as editing and managing the cells in your flow.

The following screenshot shows you the flow toolbar that is present on the main page:

Figure 2.3 – The Flow Name and flow toolbar sections

Cells are lines of code executions that you can perform one by one. They can also be used to make comments, headings, and even Markdown text.

The tools are listed as follows. We will start from left to right:

- **New**: This creates a new Flow notebook.
- **Open**: This opens an existing Flow notebook that is stored in your system.
- **Save**: This saves your current flow.
- **Insert Cell**: This inserts a cell below.
- **Move up**: This moves the currently highlighted cell up over the above cell.
- **Move down**: This moves the currently highlighted cell to below the cell under it.
- **Cut**: This cuts the cell and stores it on the clipboard.
- **Copy**: This copies the currently highlighted cell.
- **Paste**: This pastes the previously copied cell below the currently highlighted cell.
- **Clear Cell**: This clears the output of the cell if there is any.
- **Delete Cell**: This deletes the cell entirely.
- **Run and Select below**: This runs the currently highlighted cell and stops at the cell below.
- **Run**: This runs the currently highlighted cell.

- **Run all**: This runs all the cells from top to bottom one by one.

- **Assist me**: This executes an assist command that helps you by showing links to basic ML commands.

On the right-hand side of your flow are additional support options to help you manage your workflow. You have **OUTLINE**, **FLOW**, **CLIPS**, and **HELP**. Each option has its own column with functional details.

The following screenshot shows you the various support options of the H2O Flow UI:

OUTLINE   FLOWS   CLIPS   HELP

Figure 2.4 – Support options

The **OUTLINE** column shows you a list of all the executions you performed. Looking at the outline helps you to quickly check whether all the steps you performed when creating your ML pipeline were in the correct sequence, whether there were any duplicates, or whether any incorrect steps were made.

The following screenshot shows you the **OUTLINE** support option section:

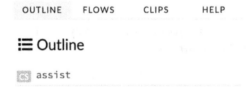

Figure 2.5 – OUTLINE

The **FLOW** column shows you all your previously saved flows. At the moment, you won't have any, but once you do save a flow, it will show up here with a quick link to open it. Please note that only one Flow notebook can be opened at a time on the H2O Flow page. If you want to work on multiple flows simultaneously, you can do so by opening another tab and opening the Flow notebook there.

The following screenshot shows you the **FLOWS** support option section:

Figure 2.6 – Saved flows

The **CLIPS** column is your clipboard. If there is any code to be executed that you will be running multiple times in your flow, then you can write it once and save it to your clipboard by selecting the paper clip icon on the right-hand side of your cell. Then, you can paste that cell whenever you need it again without needing to search for it in your flow. The clipboard also stores a set number of trash cells.

The following screenshot shows you the **CLIPS** support option section:

Figure 2.7 – Clipboard

The **HELP** column shows you various resources to help the user use H2O Flow. This involves QuickStart videos, example flows, general usage examples, and links to the official documentation. Feel free to explore these to gain an understanding of how you can use H2O Flow.

The following screenshot shows you the **Help** support option section:

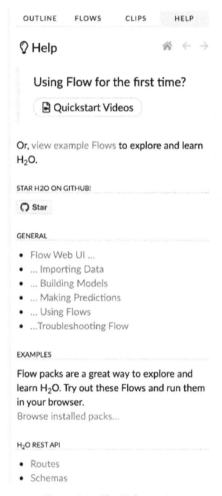

Figure 2.8 – The Help section

Back to the top of the page, let's explore the first two drop-down lists. They are simple enough to understand.

The following screenshot shows you the first drop-down list of operations, that is, the Flow object operations:

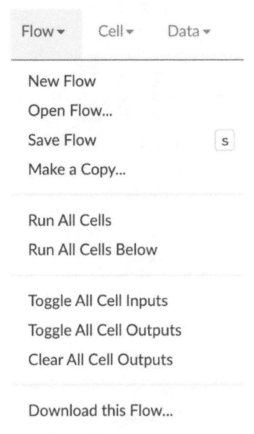

Figure 2.9 – The Flow functions drop-down list

The **Flow** drop-down list of operations has basic functionality in terms of the Flow notebook in general. You can create a new flow, open an existing one, save, copy, run all of the cells in the flow, clear the outputs, and download flows. Downloading flows is very useful, as you can easily transfer your ML work from the flows to other systems.

On the right-hand side is the **Cell** object operation drop-down list (as shown in *Figure 2.10*). This drop-down list is also simple to understand. It has basic functionality to manipulate the various cells in the flow. Usually, it targets the highlighted cell in the Flow notebook. You can change the highlighted cell by just clicking on it:

Figure 2.10 – The Cell functions drop-down list

We will explore the other options as we start working on our ML pipeline.

In this section, we understood what H2O Flow is, how to download it, and how to launch it locally. Additionally, we explored the various parts of the H2O Flow UI and the various functionalities it has to offer. Now that we are familiar with our environment, let's dive in deeper and create our ML pipeline. We will start with the very first part of the pipeline, that is, handling data.

Working with data functions in H2O Flow

An ML pipeline always starts with data. The amount of data you collect and the quality of that data play a very crucial role when training models of the highest quality. If one part of the data has no relationship with another part of the data, or if there is a lot of noisy data that does not contribute to the said relationship, the quality of the model will degrade accordingly. Therefore, before training any models, often, we perform several processes on the data before sending it to model training. H2O Flow provides interfaces for all of these processes in its **Data** operation drop-down list.

We will understand the various data operations and what the output looks like in a step-by-step process as we build our ML pipeline using H2O Flow.

So, let's begin creating our ML pipeline by, first, importing a dataset.

Importing the dataset

The dataset we will be working with in this chapter will be the `Heart Failure Prediction` dataset. You can find the details of the Heart Failure Prediction dataset at `https://www.kaggle.com/fedesoriano/heart-failure-prediction`.

This is a dataset that contains certain health information about individuals with cholesterol, maximum heart rate, chest pain, and other attributes that are important indicators of whether a person is likely to experience heart failure or not.

Let's understand the contents of the dataset:

- **Age**: The age of the patient in years
- **Sex**: The sex of the patient; M for male and F for female
- **ChestPainType**:
  - The types are listed as follows:
    - **TA**: Typical Angina
    - **ATA**: Atypical Angina
    - **NAP**: Non-Anginal Pain
    - **ASY**: Asymptomatic
- **RestingBP**: Resting blood pressure in [mm Hg]
- **Cholesterol**: Serum cholesterol in [mm/dl]
- **FastingBS**: Fasting blood sugar, where it is 1 if FastingBS is greater than 120 mg/dl, else it is 0
- **RestingECG**: Resting electrocardiogram results

- The types are listed as follows:

  - **Normal**: Normal

  - **ST**: Having ST-T wave abnormality

  - **LVH**: Showing probable or definite left ventricular hypertrophy by Estes' criteria

- **MaxHR**: The maximum heart rate achieved value, lying between 60 and 202

- **ExerciseAngina**: Exercise-induced angina; Y for yes, and N for no

- **Oldpeak**: ST (Numeric value measured in depression)

- **ST_Slope**: The slope of the peak exercise ST segment

  - The types are as follows:

    - **Up**: Sloping upward

    - **Flat**: Flat

    - **Down**: Sloping downward

- **HeartDisease**: The output class, where 1 indicates heart disease and 0 indicates no heart disease

We will train a model that tries to predict whether a person with certain attributes has the potential to face heart failure or not by using this dataset. Perform the following steps:

1. First, let's start by importing this dataset using H2O Flow.

2. On the topmost part of the web UI, you can see the **Data** object operations drop-down list.

3. Click on it, and the web UI should display an output that looks like the following screenshot:

Figure 2.11 – The Data object drop-down list

The drop-down list shows you a list of all the various operations that you can perform, along with those that are related to the data you will be working with.

The functions are listed as follows:

- **Import Files**: This operation imports files stored in your system. They can be any readable files edited in **Comma-Separated Values (CSV)** format or Excel format.

- **Import SQL Table**: This operation lets you import the data stored in your **Structured Query Language (SQL)** table if you are using an SQL database to store any tabular data. When you click on it, you will be prompted to input the **Connection URL** value to your SQL database, along with its credentials, the **Table** name, the **Columns** name, and an option to select **Fetch mode**, which could be **DISTRIBUTED** or **SINGLE**, as shown in the following screenshot:

Figure 2.12 – Import SQL Table

- **Upload File**: This operation uploads a single file from your system and parses it immediately.

- **Split Frame**: This operation splits the dataframe that has already been imported and parsed by H2O Flow into multiple frames, which you can use for multiple experiments.

- **Merge Frames**: This operation merges multiple frames into one. When you click on it, you will be prompted to select the right and left frames and their respective columns. Additionally, you have the flexibility to select which rows from which frame to include in the merge, as shown in the following screenshot:

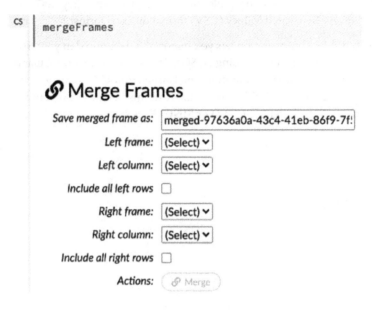

Figure 2.13 – Merge Frames

- **List frames**: This operation lists all the frames that are currently imported and parsed by H2O. This helps if you want to run your pipeline on different datasets that were previously imported or stitched by you.

- **Impute**: Imputation is the process of replacing a missing value in the dataset with an average value such that it does not introduce any bias with a minimized value, a maximized value, or even an empty value. This operation lets you replace these values in your frame depending on your preference.

So, first, let's import the `Heart Failure Prediction` dataset.

Perform the following steps:

1. Click on the **Data** operation drop-down list.

2. Click on **Import Files....** You should see an operation executed on a cell in your flow, as shown in the following screenshot:

Figure 2.14 – Import Files

You can also directly run the same command by typing `importFiles` into a cell and pressing *Shift + Enter* instead of using the drop-down list.

3. In the search bar, add the location of the folder in which you have downloaded the dataset.

4. Click on the search button on the extreme right-hand side. This will show you all the files in the folder, and you can select which ones to import.

5. Select the `heart.csv` dataset, as shown in the following screenshot:

Figure 2.15 – Importing files with inputs

6.   Once you have selected the file, click on the **Import** button. H2O will then import the dataset and show you an output, as follows:

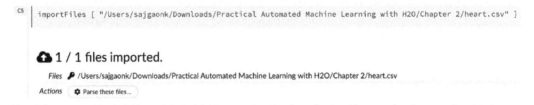

Figure 2.16 – The imported files

Congratulations! You have now successfully imported your dataset into H2O Flow. The next step is to parse it into a logical format. Let's understand what parsing and is how we can do it in H2O Flow.

Parsing the dataset

Parsing is the process of analyzing the string of information in a dataset and loading it into a logical format, in this case, a **hex** file. A hex file is a file whose contents are stored in a hexadecimal format. H2O uses this file type internally for dataframe manipulations and maintains the metadata of the dataset.

Let's parse the newly imported dataset by executing the following steps:

1.   Click on the **Parse these files…** button after importing the dataset, as shown in *Figure 2.16*.

The following screenshot shows the output you should expect:

Figure 2.17 – The dataset parsing inputs

2. Then, you will be prompted to input the **PARSE CONFIFUGRATION** settings to parse the dataset. H2O will read the dataset using these configurations and create a `hex` file. During parsing, you have the following configurations to select from:

- **Sources**: This configuration denotes the source of the dataset you will be parsing.

- **ID**: This configuration indicates the ID of the dataset. You can change the ID of the hex file that will be created.

- **Parser**: This configuration sets the parser to use for parsing your dataset. You will be using different parsers to parse different types of datasets. For example, if you are parsing a Parquet file, then you will need to select the Parquet file parser. In our case, we are parsing a CSV file; hence, select the **CSV** parser. You can set it to **AUTO** too, and H2O will self-identify which parser will be needed to parse the specified file.

- **Separator**: This configuration is used to specify the separator that separates the values in the dataset. Let's leave it at the default value, **,:'004'**, which is a very commonly found CSV separator value.

- **Escape Character**: This configuration is used to specify the escape character that is used in the dataset. Let's leave it at the default value.

- **Column Headers**: This configuration is used to specify whether the first row of the dataset contains column names or not. H2O will use this information to handle the first row to either use it as column information or autogenerate column names and use the first row as data values. Additionally, you can set the value to **Auto** to let H2O self-identify whether the first row of the dataset contains column names or not. Our dataset has the first row as column names, so let's select **First row contains column names**.

- **Options**: This configuration contains additional operations as follows:

  - **Enable single quotes as a field quotation character**: This option enables single quotes as a field quotation character.

  - **Delete on done**: This option deletes the imported dataset once it has been successfully parsed. Let's tick mark this option, as we won't be needing the imported dataset once we successfully parse it.

3. Below the **PARSE CONFIGURATION** section, you have the **Edit Column names and Types** section (see *Figure 2.18*). Before parsing the dataset, H2O reads the dataset and shows you a brief summary of the column names and the types of information they extracted. This gives you the option of editing your column names and types in case it has interpreted something incorrectly or if you wish to change the column names to something different.

The following screenshot shows you the column names and the option to edit their type when parsing the dataset:

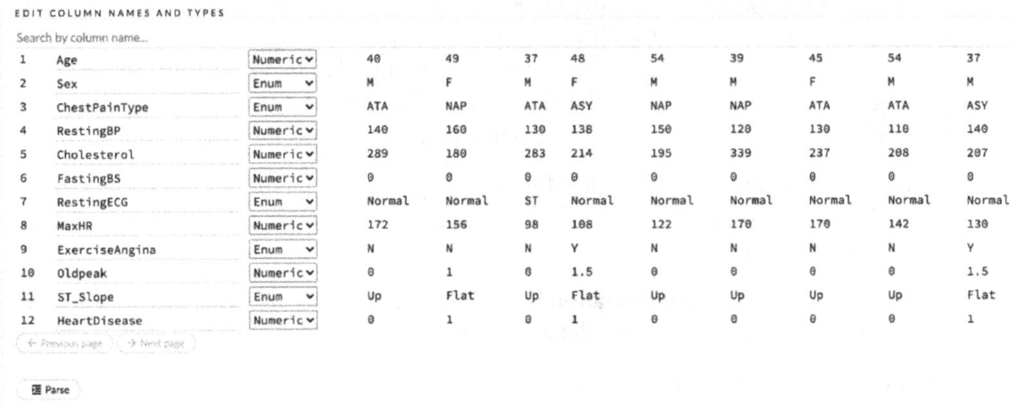

EDIT COLUMN NAMES AND TYPES

Search by column name...

1	Age	Numeric ∨	40	49	37	48	54	39	45	54	37
2	Sex	Enum ∨	M	F	M	F	M	M	F	M	M
3	ChestPainType	Enum ∨	ATA	NAP	ATA	ASY	NAP	NAP	ATA	ATA	ASY
4	RestingBP	Numeric ∨	140	160	130	138	150	120	130	110	140
5	Cholesterol	Numeric ∨	289	180	283	214	195	339	237	208	207
6	FastingBS	Numeric ∨	0	0	0	0	0	0	0	0	0
7	RestingECG	Enum ∨	Normal	Normal	ST	Normal	Normal	Normal	Normal	Normal	Normal
8	MaxHR	Numeric ∨	172	156	98	108	122	170	170	142	130
9	ExerciseAngina	Enum ∨	N	N	N	Y	N	N	N	N	Y
10	Oldpeak	Numeric ∨	0	1	0	1.5	0	0	0	0	1.5
11	ST_Slope	Enum ∨	Up	Flat	Up	Flat	Up	Up	Up	Up	Flat
12	HeartDisease	Numeric ∨	0	1	0	1	0	0	0	0	1

← Previous page   → Next page

⊞ Parse

Figure 2.18 – Editing the column types

Now, let's briefly understand the types of the columns, as shown in the preceding screenshot:

They can be listed as follows:

- **Numeric**: This indicates that the values in the column are numbers.

- **Enum**: This indicates that the values in the column are **Enumerated Types** (**Enums**). Enums are a certain set of named values that are non-numeric and finite in nature.

- **Time**: This indicates that the values in the column are datetime data type values.

- **UUID**: This indicates that the values in the column are **Universally Unique Identifiers** (**UUIDs**).

- **String**: This indicates that the values in the column are string values.

- **Invalid**: This indicates that the values in the column are invalid, which means that there are some issues with the values in the column.

- **Unknown**: This indicates that the values in the column are unknown, which means that the column contains no values.

During training, H2O treats the columns of the dataframe differently depending on their type. Sometimes, it misinterprets the values of the columns and assigns an incorrect column type to a column. Take a look at the columns of the datasets and their corresponding types. You might notice that after importing the dataset, the type of our **HeartDisease** output column is **Numeric**. H2O read the 1 and 0 and assumed its type was Numeric. But it is, in fact, an **Enum** value, as 1 indicates heart disease, and 0 indicates no heart disease. There is no numerical value in between. We always need to be careful when parsing datasets. We need to ensure that H2O interprets the dataset correctly before we start training models on it. H2O Flow provides an easy way to correct this before parsing.

Next, let's correct the column types:

1. Select the drop-down list next to the **HeartDisease** column name and set it to **Enum**.

2. Similarly, **FastingBS** will be set as **Numeric**, but it is, in fact, an **Enum** value based on its description. So, let's set it to **Enum**.

3. Now that we have corrected the column types and have selected the correct parsing configurations, let's parse the dataset. Click on the **Parse** button at the bottom of the output.

   The following screenshot shows you the output after parsing:

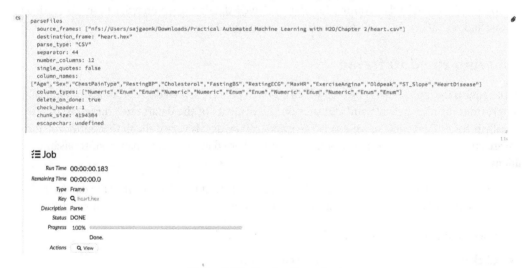

Figure 2.19 – Parsing the output

Now, let's observe the output we got from the preceding screenshot:

- **Run Time**: This output shows you the runtime since H2O started to parse the dataset.
- **Remaining Time**: This output shows you the expected remaining time it will take H2O to finish parsing the dataset.
- **Type**: This output shows you the parsed file's type.
- **Key**: This output shows you a link to the hex file, which has been generated after successfully parsing.
- **Description**: This output shows you the operation currently being performed.
- **Status**: This output shows you the current status of the operation.
- **Progress**: This output shows you the progress bar of the operation.
- **Action**: This output is a button that shows you the hex file that has been generated after successfully parsing the dataset.

Congratulations! You have successfully parsed your dataset and have generated the hex file. The hex file is commonly termed a **dataframe**. Now that we have a dataframe generated, let's see the various metadata and operations available to us to be performed on the dataframe.

Observing the dataframe

The dataframe is the primary data object on which H2O can perform several data operations. Additionally, H2O provides detailed insights and statistics on the contents of the dataframe. These insights are especially helpful when working with very large dataframes, as it is very difficult to identify whether there are any missing or zeros in any of the columns where data rows can span from thousands to millions.

So, let's observe the dataframe we just parsed and explore its features. You can view the dataframe by performing either of the following actions from the parsing output that we saw in *Figure 2.19*:

- Click on the `heart.hex` hyperlink in the **Key** row.
- Click on the **View Data** button in the **Actions** row.

The following screenshot shows you the output of the previously mentioned actions:

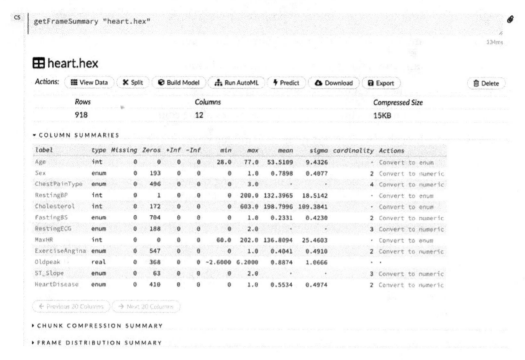

Figure 2.20 – Viewing a dataframe

The output displays a summary of the dataframe. Before exploring the various operations in the **Actions** section, first, let's explore the metadata of the dataframe that is shown below it.

The metadata consists of the following:

- **Rows**: Displays the number of rows in the dataset
- **Columns**: Displays the number of columns in the dataset
- **Compressed Size**: Displays the total compressed size of the dataframe

Below the metadata, you have the **COLUMN SUMMARIES** section. Here, you can see statistics about the contents of the dataframe split across the columns.

The summary contains the following information:

- **label**: This column shows the name of the columns of the dataframe.
- **type**: This column shows the type of the column.
- **Missing**: This column shows the number of missing values in that column.

- **Zeros**: This column shows the number of zeros in the column.

- **+Inf**: This column shows the number of positive infinity values.

- **-Inf**: This column shows the number of negative infinity values.

- **min**: This column shows the minimum value in the column.

- **max**: This shows the maximum value in the column.

- **sigma**: This column shows the variability in the values of the column.

- **cardinality**: This column shows the number of unique values in the column.

- **Actions**: This column shows certain actions that you can perform on the columns of the dataframe. These actions mostly consist of converting the columns into different types. It is useful in correcting the column types if they were read incorrectly by H2O after parsing.

You might be wondering why you are seeing plenty of values in the **Zeros** columns for columns such as **Sex** and **RestingECG**. This is because of **encoding**. When parsing enum columns, H2O encodes enums and strings into numerical values starting from 0 to 1, then 2, and so on and so forth. You can also control this encoding process by selecting which encoding process you want, for example, **Label encoding** or **One-hot encoding**. We will discuss encoding in greater detail in *Chapter 5, Understanding AutoML Algorithms*.

After the **COLUMN SUMMARIES** section, you will see the **CHUNK COMPRESSION SUMMARY** section, as shown in the following screenshot:

▾ CHUNK COMPRESSION SUMMARY

chunk_type	chunk_name	count	count_percentage	size	size_percentage
CBS	Binary	4	33.3333	740 B	7.7390
C1N	1-Byte Integers (w/o NAs)	6	50.0	5.8 KB	61.8699
C1S	1-Byte Fractions	1	8.3333	1002 B	10.4790
C2	2-Byte Integers	1	8.3333	1.9 KB	19.9122

Figure 2.21 – The dataset's CHUNK COMPRESSION SUMMARY section

When working with very large datasets, it can take a very long time to read and write data from the datasets if done in a traditional manner. Therefore, systems such as Hadoop File System, Spark, and more are often used as they perform read and write in a distributed manner, which is faster than the traditional contiguous manner. **Chunking** is a process where distributed file systems such as Hadoop split the dataset into chunks and flatten them before writing to disk. H2O handles all of this internally so that users do not need to worry about managing distributed read or write. The **CHUNK COMPRESSION SUMMARY** section just gives additional information about the chunk distribution.

Underneath **CHUNK COMPRESSION SUMMARY** is the **FRAME DISTRIBUTION SUMMARY** section, as shown in the following screenshot:

▾ FRAME DISTRIBUTION SUMMARY

	size	number_of_rows	number_of_chunks_per_column	number_of_chunks
192.168.0.157:54321	9.3 KB	918.0	1.0	12.0
mean	9.3 KB	918.0	1.0	12.0
min	9.3 KB	918.0	1.0	12.0
max	9.3 KB	918.0	1.0	12.0
stddev	0 B	0	0	0
total	9.3 KB	918.0	1.0	12.0

Figure 2.22 – The dataset's Frame Distribution Summary section

The **FRAME DISTRIBUTION SUMMARY** section gives statistical information about the dataframe in general. If you have imported multiple files and parsed them into a single dataframe, then the frame distribution summary will calculate the total size, the mean size, the minimum size, the maximum size, the number of chunks per dataset, and more.

Looking back at the dataframe from *Figure 2.20*, you can see that the column names are highlighted as links. Let's click on the last column, **HeartDisease**.

The following screenshot will show you a column summary of the **HeartDisease** column:

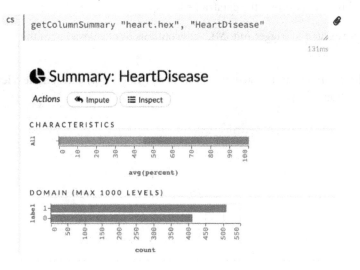

Figure 2.23 – Column summary

The **Column Summary** gives you more details about individual columns in the dataset based on their type.

For enums such as **HeartDisease**, it shows you the details of the column as follows:

- **Characteristics**: This graph shows you the percentage distribution of zero and nonzero values.
- **Domain**: This graph shows you the domain of the column, that is, the number of times a value has been repeated over the total number of rows.

You will see more details when you hover your mouse over the graphs. Feel free to explore all the columns and try to understand their features.

Returning to the dataframe from *Figure 2.20*, let's now look at the **Actions** section. This includes certain actions that are available to be performed on the dataframe.

Those actions include the following:

- **View Data**: This action displays the data in the dataframe.
- **Split**: This action splits the dataframe into the specified parts in the specified ratio.
- **Build Model**: This action starts building a model on this dataframe.
- **Run AutoML**: This action triggers AutoML on the dataframe.
- **Predict**: This action uses the dataframe to predict on an already trained model and gets results.
- **Download**: This action downloads the dataframe to the system.
- **Export**: This action exports the dataframe to a specified path in the system.

As much as we are excited to trigger AutoML, let's continue our exploration of the various dataframe features.

Now, let's see the dataframe and its actual data contents. To do this, click on the **View Data** button in the **Actions** section.

The following screenshot shows you the output of the **View Data** button:

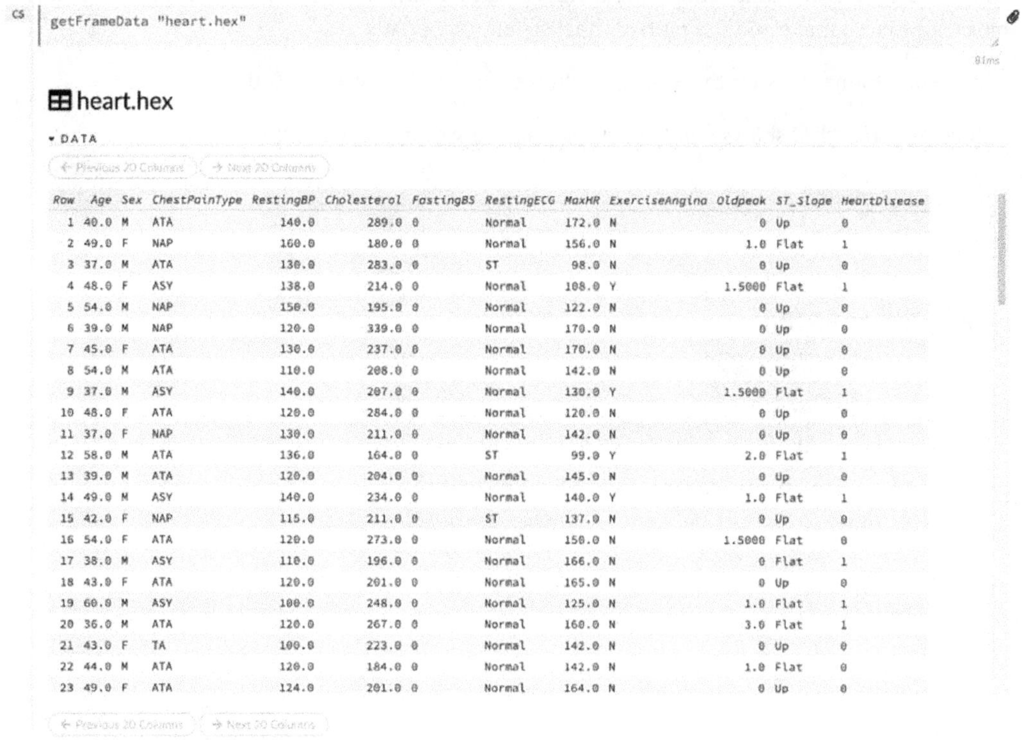

CS `getFrameData "heart.hex"`

81ms

⊞ heart.hex

▾ DATA

(← Previous 20 Columns) (→ Next 20 Columns)

Row	Age	Sex	ChestPainType	RestingBP	Cholesterol	FastingBS	RestingECG	MaxHR	ExerciseAngina	Oldpeak	ST_Slope	HeartDisease
1	40.0	M	ATA	140.0	289.0	0	Normal	172.0	N	0	Up	0
2	49.0	F	NAP	160.0	180.0	0	Normal	156.0	N	1.0	Flat	1
3	37.0	M	ATA	130.0	283.0	0	ST	98.0	N	0	Up	0
4	48.0	F	ASY	138.0	214.0	0	Normal	108.0	Y	1.5000	Flat	1
5	54.0	M	NAP	150.0	195.0	0	Normal	122.0	N	0	Up	0
6	39.0	M	NAP	120.0	339.0	0	Normal	170.0	N	0	Up	0
7	45.0	F	ATA	130.0	237.0	0	Normal	170.0	N	0	Up	0
8	54.0	M	ATA	110.0	208.0	0	Normal	142.0	N	0	Up	0
9	37.0	M	ASY	140.0	207.0	0	Normal	130.0	Y	1.5000	Flat	1
10	48.0	F	ATA	120.0	284.0	0	Normal	120.0	N	0	Up	0
11	37.0	F	NAP	130.0	211.0	0	Normal	142.0	N	0	Up	0
12	58.0	M	ATA	136.0	164.0	0	ST	99.0	Y	2.0	Flat	1
13	39.0	M	ATA	120.0	204.0	0	Normal	145.0	N	0	Up	0
14	49.0	M	ASY	140.0	234.0	0	Normal	140.0	Y	1.0	Flat	1
15	42.0	F	NAP	115.0	211.0	0	ST	137.0	N	0	Up	0
16	54.0	F	ATA	120.0	273.0	0	Normal	150.0	N	1.5000	Flat	0
17	38.0	M	ASY	110.0	196.0	0	Normal	166.0	N	0	Flat	1
18	43.0	F	ATA	120.0	201.0	0	Normal	165.0	N	0	Up	0
19	60.0	M	ASY	100.0	248.0	0	Normal	125.0	N	1.0	Flat	1
20	36.0	M	ATA	120.0	267.0	0	Normal	160.0	N	3.0	Flat	1
21	43.0	F	TA	100.0	223.0	0	Normal	142.0	N	0	Up	0
22	44.0	M	ATA	120.0	184.0	0	Normal	142.0	N	1.0	Flat	0
23	49.0	F	ATA	124.0	201.0	0	Normal	164.0	N	0	Up	0

(← Previous 20 Columns) (→ Next 20 Columns)

Figure 2.24 – The dataframe's content view

This function shows the actual contents of the dataset. You can scroll down to see all of the contents of the dataframe. If you have run any operations on the dataframe, then you will be able to view the changes in the content of the dataframe using this operation.

Now that you have explored and understood the various dataframe operations and metadata of the dataframe, let's investigate another interesting dataframe operation called **Split**.

Splitting a dataframe

Before we send a dataframe for model training, we need a sample dataframe with predicted values. This sample can be used to validate whether the model is making a correct prediction or not and measure the accuracy and performance of the model. We can do this by reserving a small part of the dataframe to be used later for validation. This is where split functionality comes into the picture.

Split, as the name suggests, splits the dataset into parts that we can later use for different operations. Splitting of dataset takes ratios into account. H2O creates multiple dataframes based on the number of splits you want to make and the ratio of the distribution of data.

To split the dataframe, click on the **Split** operation button in the **Actions** section.

The following screenshot shows you the output of clicking on the **Split** button:

Figure 2.25 – Splitting a dataframe

In the preceding output, you will be prompted to input the splitting configurations, which are listed as follows:

- **Frame**: This configuration lets you select the frame that you want to split.
- **Splits**: This configuration lets you select the number of splits that you want to make and the ratio of the distribution of data you want among them. Additionally, you can set the key names for the splits. For our walk-through, let's split the dataframe into three dataframes:

  - `training_dataframe` with a ratio of 0.70
  - `validation_dataframe` with a ratio of 0.20
  - `testing_dataframe` with a ratio of 0.10

- **Seed**: This configuration lets you select the randomization point. Split does not split the dataframe linearly. It splits the rows of the dataframe randomly. This is a good thing, as it equally distributes the data between the split dataframes, thus removing any bias that might be present in the sequence of the dataset rows. For our walk-through, let's set the seed value to 5.

Once you have made these changes, click on the **Create** button. This will generate three dataframes: **training_dataframe**, **validation_dataframe**, and **testing_dataframe**. These three frames will be stored in H2O's local repository and can be available for other experiments, too.

The following diagram shows you the output of the **Create** operation:

Figure 2.26 – Split dataset results

All three of the dataframes are independent dataframes and have the same set of features and options as you saw for the original dataset in *Figure 2.20*. Splitting does not delete the original dataframe. It creates new dataframes and copies the data from the original dataframe and distributes them among the dataframes based on the selected ratios.

Now that we have the training frame, the validation frame, and the testing frame ready, we are ready for the next step of the model training pipeline, that is, model training.

So, to recap, in this section, you understood the various data operations we can perform on the dataset available in H2O Flow such as importing, parsing, reading the parsed dataframe, and splitting it.

In the next section, we will focus on model training and using H2O AutoML to train models on the dataframes we created in this heading.

Working with model training functions in H2O Flow

Once your dataset is ready, the next step of the model training pipeline is the actual training part. The training of models can get very complex as there are a lot of configurations that decide how the model will be trained on the dataset. This is true even for AutoML where the majority of the hyperparameter tuning is done behind the scenes. Not only are there right and wrong ways of training a model for a specific type of data, but some of the configuration values can also affect the performance of the model. Therefore, it is important to understand the various configuration parameters that H2O has to offer when training a model using AutoML. In this section, we will focus on understanding what these parameters are and what they do when it comes to model training.

We will understand how to train a model using AutoML, step by step, using the dataframes we created previously.

Note that there are plenty of things in this section that will be too complex to understand at this stage. We will explore some of them in future chapters. For the time being, we will only focus on features we can understand right now, as the goal of this chapter is to understand H2O Flow and how to use AutoML to train models using H2O Flow.

In the following sub-sections, we will gain an understanding of the model training operations, starting with an understanding of the AutoML training configuration parameters.

Understanding the AutoML parameters in H2O Flow

H2O AutoML is extremely configurable in terms of how you want to train your models. Despite using the same AutoML technology, often, every industry will have certain preferences or inclinations regarding how it wants to train its models based on their requirements. So, even though AutoML is automating most of the ML processes, a degree of control and flexibility is still needed in terms of how AutoML trains its models. H2O provides extensive configuration capabilities for its AutoML feature. Let's explore them as we train our model.

There are two ways that you can trigger AutoML on a dataset. They are listed as follows:

- By clicking on the **Run AutoML** button in the **Actions** section of your dataframe output, as shown in *Figure 2.20*.

- By selecting **Run AutoML** in the drop-down menu of the **Model** section, on the topmost part of the web UI, as shown in *Figure 2.27*.

Let's go with the second option so that we can explore the **Model** operations section at the top of the web UI page. When you click on the **Model** section, you should see the drop-down list, as shown in the following screenshot:

Figure 2.27 – The Model functions drop-down list

The preceding drop-down list categorizes the model operations into three types, as follows:

- **Run AutoML**: This operation starts the AutoML process by prompting the user to input configuration values to train models using AutoML.

- **Run Specific Models**: This operation starts model training on specific ML algorithms that you wish to use; for example, deep learning, K-means clustering, random forest, and more. Each ML algorithm comes with its own set of parameters that you must specify to start training models.

- **Model Management options**: These operations are basic operations that are used to manage the various models you might have trained over a period of time.

  - The operations are listed as follows:

    - **Import MOJO Model**: This operation imports H2O **Model Object, Optimized (MOJO)** models previously trained by another H2O service and exported to the system in the form of a MOJO.

    - **List All Models**: This operation lists all the models trained by your H2O service. This includes those trained by other Flow notebooks.

    - **List Grid Search Results**: Grid search is a technique that is used to find the best hyperparameters when training a model to get the best and most accurate performance out of it. This operation lists all the grid search results from the model training.

    - **Import Model**: This operation imports model objects into H2O.

    - **Export Model**: This operation exports model objects to the system either as a binary file or as a MOJO.

Now, let's use AutoML to train our model on our dataset.

Click on **Run AutoML**. As shown in *Figure 2.27*, you should see an output prompting you to input a wide variety of configuration parameters to run AutoML. There are tons of options to configure your AutoML training. These can greatly affect your model training performance and the quality of the models that eventually get trained. The parameters are classified into three categories. Let's explore them, one by one, categorically and input the values to suit our model training requirements starting with the basic parameters.

Basic parameters

Basic parameters are parameters that focus on the basic inputs needed for all model training operations. These are common among all the ML algorithms and are self-explanatory.

The following screenshot shows you the basic parameter values you need to input to configure AutoML:

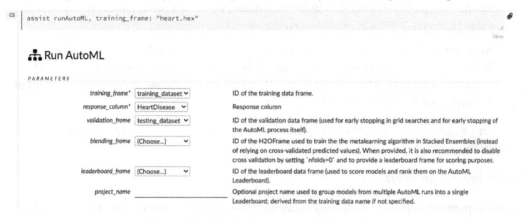

Figure 2.28 – The basic parameters of AutoML

The basic parameters are listed as follows:

- **training_frame**: This configuration sets the dataframe to be used for training the model and is mandatory. In our case, our `training_frame` parameter is the `training_dataframe.hex` file that we created in the *Working with data functions in H2O Flow* section.

- **response_column**: This configuration sets the response column of the dataframe that is to be predicted. This is a mandatory parameter. In our case, our response column is the `HeartDisease` column.

- **validation_frame**: This configuration sets the dataframe to be used for validation of the model during training. In our case, our validation frame is the `validation_dataframe.hex` file we created in the previous section.

- **blending_frame**: This configuration sets the dataframe that should be used to train the stacked ensemble models. For the time being, we can ignore this as we won't be exploring stacked ensemble very much in this chapter. We will explore stacked ensemble models in more detail in *Chapter 5, Understanding AutoML Algorithms*, so let's leave this blank.

- **leaderboard_frame**: This configuration sets the ID of the dataframe, which will be used to calculate the performance metrics of the trained models, and it will use the results to rank them on the leaderboard. If a leaderboard frame has not been specified, then AutoML will use cross-validation metrics to rank the models. Even if cross-validation is turned off by setting `nfolds` to 0, then AutoML will generate a leaderboard from the training frame.

- **project_name**: This configuration sets the name of your project. AutoML will group all of the results from multiple runs into a single leaderboard under this project name. If you leave this value blank, H2O will autogenerate a random name for the project on its own.

- **distribution**: This configuration is used to specify the type of distribution function to be used by ML algorithms. Algorithms that support the specified type of distribution function will use it, while others will use their default value. We will learn more about distribution functions in *Chapter 5, Understanding AutoML Algorithms*.

Now that you understood what the basic parameters of AutoML are, let's gain an understanding of the advanced parameters.

Advanced parameters

Advanced parameters provide additional configurations for training models using AutoML. These parameters have certain default values set to them and, therefore, are not mandatory. However, they do provide additional configurations that change the behavior of AutoML when training models.

The following screenshots show you the first part of the advanced parameters for AutoML:

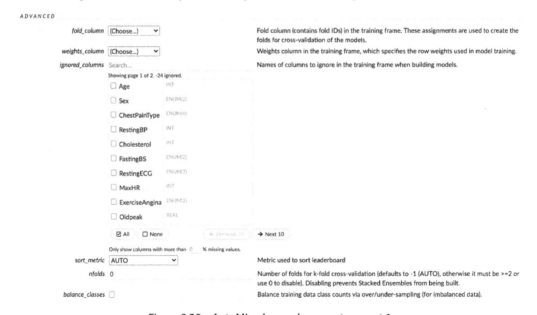

Figure 2.29 – AutoML advanced parameters, part 1

Immediately below this will be the second part of the advanced parameters. Scroll down to see the remaining parameters, as shown in the following screenshot:

Figure 2.30 – AutoML advanced parameters, part 2

Let's explore the parameters one by one. The advanced parameters are listed as follows:

- **fold_column**: This parameter uses the column values as a basis to create its folds in N-Fold cross-validation. We will learn more about cross-validation in the upcoming chapters.

- **weights_column**: This parameter gives weight to a column in the training frame or, put simply, gives preference to a column. We will explore weights in model training in greater detail in later chapters.

- **ignored_columns**: This parameter lets you select the columns from the training frame that you wish to ignore when training models. We won't be ignoring any columns for our experiment, so we will leave all the columns unchecked.

- **sort_metrics**: This parameter selects the metric that will be used to sort and rank the models trained by AutoML. Selecting **AUTO** will use the **AUC** metrics for binary classification, **mean_per_class_error** for multinomial classification, and **deviance** for regression. For simplicity's sake, let's set **Mean Squared Error** (**MSE**) as the sorting metric since it is easier to understand. We will explore metrics in greater detail in *Chapter 6, Understanding H2O AutoML Leaderboard and Other Performance Metrics*.

- **nfolds**: This parameter selects the number of cross-validation folds created for cross-validation. Cross-validation generates stacked ensemble models. For the moment, we will set this value to 0 to prevent the creation of stacked ensemble models. We will explore stacked ensemble models and cross-validation further in the upcoming chapters.

- **Balance_classes**: In model training, it is always best to train models on datasets that have output classes with an equal distribution of values. This mitigates any biases that might arise from an unequal distribution of values. This parameter balances the output classes to be of equal numbers. In our case, the output class is the **Heart Disease** column. As we saw in the column summary of the **Heart Disease** column, in the distribution section, we had around 410 values as 0 and 508 values as 1 (*see Figure 2.23*). So, ideally, we need to balance out the classes before training a model. However, to keep things simple for the moment, we will skip the balancing of the classes and focus on understanding the basics. We will explore class balancing further in the upcoming chapters.

- **exclude_algos**: This parameter excludes certain algorithms from the AutoML training. For our experiment, we want to consider all the algorithms, so let's leave the value empty.

- **exploitation_ratio**: This is an experimental option that sets the exploitation versus exploration ratio when training a model. We will discuss this in greater detail in the upcoming chapters. Let's keep the default value of -1 as it is.

- **monotone_constraints**: In some cases, where there is a very strong prior belief that the relationship between features has some quality, constraints can be used to improve the predictive performance of the model. This parameter helps set these constraints. We will discuss this more in the upcoming chapters. Let's leave this value empty.

- **seed**: This parameter sets the seed value for randomization, which is useful for reproducing results. Let's set this to 5.

- **max_models**: This parameter sets the maximum number of models to be trained by H2O AutoML. We will set this to 10, else model training can take a long time.

- **max_runtime_secs**: This parameter sets the maximum time that H20 should spend training a single model. Usually, the larger the size of the dataset, the more time it takes for a model to train. Setting a very small runtime for model training will not give AutoML the time it needs to train models. Since our dataset is not very big, we will let AutoML take its time training the models, which shouldn't be that long.

- **max_runtime_sec_per_model**: This parameter sets the maximum time AutoML should spend on training an individual model. Model training won't take too long. So, let's ignore this.

- **stopping_round**: This tells AutoML to stop training models after the stopping metric doesn't improve much over the number of rounds. We want to train all the models, so we will set this value to 0 to disable it.

- **stopping_metrics**: This metric is used for early stopping. Since stopping is disabled, we will ignore this.

- **stopping_tolerance**: This refers to the relative tolerance of improvement on progressive model training expected below which AutoML should stop training models. Since stopping is disabled, we will ignore this.

Now that you have understood what the basic parameters of AutoML are, let's check the expert parameters.

Expert parameters

Expert parameters are parameters that provide additional options that supplement the AutoML training results with additional features that help in further experimentation. These parameters are dependent on the configuration you already selected in the *Advanced parameters* section.

The following screenshot shows you the expert parameter options that are available for our current walk-through:

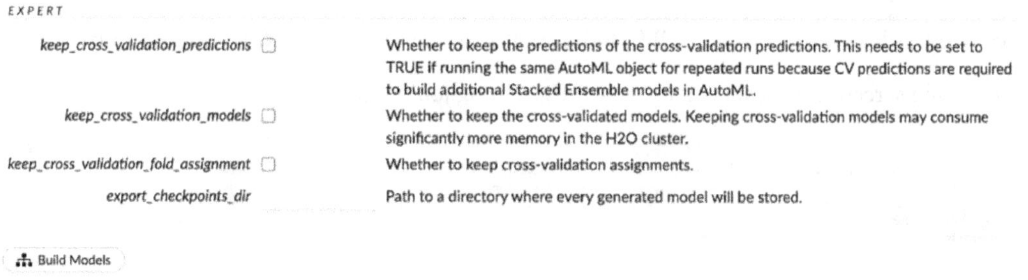

EXPERT

keep_cross_validation_predictions ☐	Whether to keep the predictions of the cross-validation predictions. This needs to be set to TRUE if running the same AutoML object for repeated runs because CV predictions are required to build additional Stacked Ensemble models in AutoML.
keep_cross_validation_models ☐	Whether to keep the cross-validated models. Keeping cross-validation models may consume significantly more memory in the H2O cluster.
keep_cross_validation_fold_assignment ☐	Whether to keep cross-validation assignments.
export_checkpoints_dir	Path to a directory where every generated model will be stored.

🔀 Build Models

Figure 2.31 – AutoML expert parameters

The expert parameters are listed as follows:

- **keep_cross_validation_predictions**: If cross-validation is enabled for training, then H2O provides the option of saving those prediction values.

- **keep_cross_validation_models**: If cross-validation is enabled for training, then H2O provides the option of saving the models trained for the same.

- **keep_cross_validation_fold_assignments**: If cross-validation is enabled for training, then H2O provides the option of saving the folds used in model training for the different cross-validations.

- **export_checkpoint_dir**: This is the directory path where H2O will store the generated models.

More expert options become available depending on what you select in the basic and advanced parameters. For the time being, we can disable all the expert parameters as we won't be focusing much on them in this walk-through.

Once you have set all of the parameter values, the only thing left now is to trigger the AutoML model training.

Training and understanding models using AutoML in H2O Flow

Model training is one of the most complex and important parts of the ML pipeline. Model training is the process of mapping a mathematical approximation by learning the relationship between features and the expected output, all while trying to minimize loss. There are various ways you can do this. The method by which a system performs this task is known as an ML algorithm. AutoML trains models using various ML algorithms and compares their performance with each other to find the one that has the least error metric value as per the ML problem.

First, let's gain an understanding of how we can train a model using AutoML in H2O Flow.

Training models using AutoML in H2O Flow

You need to make careful considerations when setting the parameter values for training models using AutoML. Once you have selected the correct inputs, you can then trigger the AutoML model training.

You can do this by clicking on the **Build Models** button at the end of your **Run AutoML** output.

The following screenshot shows you the AutoML training job progress:

```
runAutoML {"input_spec":
{"training_frame":"training_dataset","response_column":"HeartDisease","validation_frame":"testing_dataset","ignored_columns":
[],"sort_metric":"AUTO"},"build_control":{"nfolds":0,"balance_classes":false,"stopping_criteria":
{"seed":5,"max_models":10,"max_runtime_secs":0,"max_runtime_secs_per_model":0,"stopping_rounds":3,"stopping_metric":"AUTO","stopping_tolerance":-1},
"keep_cross_validation_predictions":false,"keep_cross_validation_models":false,"keep_cross_validation_fold_assignment":false},"build_models":
{"exclude_algos":[],"exploitation_ratio":-1,"monotone_constraints":[]}}, 'exec'
```

:≡ Job

Run Time	00:00:10.158
Remaining Time	00:00:00.0
Type	Auto Model
Key	🔍 AutoML_1_20220104_223006@@HeartDisease
Description	AutoML build
Status	DONE
Progress	100%
	Done.
Actions	🔍 View

Figure 2.32 – The AutoML training is finished

It should take some time to train the models. The results of the model training are stored on a **Leaderboard**. The value of the **Key** section is a link to the leaderboard (see *Figure 2.32*).

You can view the leaderboard in one of two ways. They are listed as follows:

- Click on the leaderboard link in the **Key** section of the AutoML training job output.
- Click on the **View** button of the AutoML training job output.

In the following screenshot, you can see what a **Leaderboard** looks like:

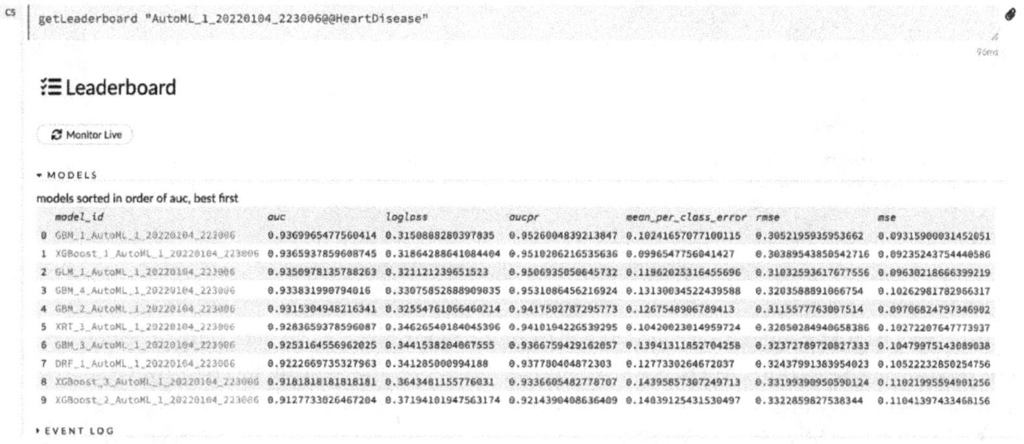

	model_id	auc	logloss	aucpr	mean_per_class_error	rmse	mse
0	GBM_1_AutoML_1_20220104_223006	0.9369965477560414	0.3150888280397835	0.9526004839213847	0.10241657077100115	0.3052195935953662	0.09315900031452051
1	XGBoost_1_AutoML_1_20220104_223006	0.9365937859608745	0.31864288641084404	0.9510206216535636	0.0996547756041427	0.30389543850542716	0.09235243754440586
2	GLM_1_AutoML_1_20220104_223006	0.9350978135788263	0.321121239651523	0.9506935050645732	0.11962025316455696	0.31032593617677556	0.09630218666399219
3	GBM_4_AutoML_1_20220104_223006	0.933831990794016	0.33075852688909035	0.9531086456216924	0.13130034522439588	0.3203588891066754	0.10262981782966317
4	GBM_2_AutoML_1_20220104_223006	0.9315304948216341	0.32554761066460214	0.9417502787295773	0.1267548906789413	0.3115577763007514	0.09706824797346902
5	XRT_1_AutoML_1_20220104_223006	0.9283659378596087	0.34626540184045396	0.9410194226539295	0.10420023014959724	0.32050284940658386	0.10272207647773937
6	GBM_3_AutoML_1_20220104_223006	0.9253164556962025	0.3441538204067555	0.9366759429162057	0.13941311852704258	0.32372789720827333	0.10479975143089038
7	DRF_1_AutoML_1_20220104_223006	0.9222669735327963	0.341285000994188	0.937780404872303	0.1277330264672037	0.32437991383954023	0.10522232850254756
8	XGBoost_3_AutoML_1_20220104_223006	0.9181818181818181	0.3643481155776031	0.9336605482778707	0.14395857307249713	0.33199390950590124	0.11021995594901256
9	XGBoost_2_AutoML_1_20220104_223006	0.9127733026467204	0.37194101947563174	0.921439040863640 9	0.14039125431530497	0.3322859827538344	0.11041397433468156

Figure 2.33 – AutoML Leaderboard

If you followed the same steps, as shown in the previous examples, then you should see the same output, albeit with slightly different IDs for the models as it is randomly generated.The leaderboard shows you all the models that AutoML has trained and ranks them from best to worst based on the **sorting metric**. A sorting metric is a statistical measurement of the quality of a model, which is used to compare the performance of different models. Additionally, the leaderboard has links to all the individual model details. You can click on any of them to get more information about the individual models. If AutoML is in progress and currently training models, then you can also view the progress of the training in real time by clicking on the **Monitor Live** button.

Now that you have understood how to train a model using AutoML, let's dive deep into the ML model details and try to understand its various characteristics.

Understanding ML models

An ML model can be described as an object that contains a mathematical equation that can identify patterns for a given set of features and predict a potential outcome. These ML models form the central component of all ML pipelines, as the entire ML pipeline aims to create and use these models for predictions. Therefore, it is important to understand the various details of a trained ML model in order to justify whether the predictions that it is making are accurate or not and to what degree they are accurate.

Let's click on the best model on the leaderboard to understand more of its details. You should see the following output:

Figure 2.34 – Model information

As you can see, H2O provides a very in-depth view of the model details. All the details are categorized inside their own sub-sections. There is a lot of information here, and as such, it can be overwhelming. For the moment, we will only focus on the important parts.

Let's glance through the important and easy-to-understand details one by one:

- **Model ID**: This is the ID of the model.

- **Algorithm**: This indicates the algorithm used to train the model.

- **Actions**: Once the model has been trained, you can perform the following actions on it:

  - **Refresh**: This refreshes the model in memory.

  - **Predict**: This lets you start making predictions on this model.

  - **Download POJO**: This downloads the model in **Plain Old Java Object** (**POJO**) format.

  - **Download Model Deployment Package**: This downloads the model in MOJO format.

  - **Export**: This exports the model as either a file or MOJO to the system.

  - **Inspect**: H2O goes into inspect mode, where it inspects the model object, retrieving detailed information about its schemas and sub-schemas.

  - **Delete**: This deletes the model.

  - **Download Gen Model**: This downloads the generated model JAR file.

Then, you have a set of sub-sections explaining more about the model's metadata and performance. These sub-sections vary slightly for different models, as some models might need to show some additional metadata for explainability purposes.

As you can see, a lot of the details seem very complex in nature and they are rightfully so. This is because they involve a bit of data science knowledge. Don't worry; we will explore all of them in the upcoming chapters.

Let's go through the easier sub-sections so that we can understand the various metadata of the ML models:

- **Model Parameters**: The model parameters are nothing but the input parameters we passed to AutoML, along with some inputs AutoML decided to use to train the model. You can choose to view either all of the parameters or only the modified ones by clicking on the **Show all parameters** or **Show modified parameters** buttons. The description of the parameters is shown next to every parameter.

- **Variable Importances**: Variable importances denote the importance of the variable when it comes to making predictions. Variables with the most importance are the ones the ML model relies most on when making predictions. Any changes to variables of high importance can drastically affect the model prediction.

The following screenshot shows you the scaled importance of various features from the dataframe that we used to train models:

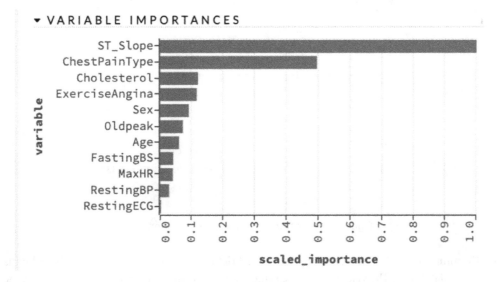

Figure 2.35 – Variable importances

- **Output**: The output sub-section denotes the basic output value of the AutoML training. The majority of the details that are shown are for cross-validation results.

The output values are listed as follows:

- **model_category**: This value indicates the category of the model that denotes what kind of prediction it performs. Binomial indicates that it performs binomial classification, which means it predicts whether the value is either **1** or **0**. In our case, **1** indicates that the person is likely to face a heart condition, and **0** indicates that the person is not likely to face a heart condition.

- **start_time**: This value indicates the epoch time at which it started the model training.

- **end_time**: This value indicates the epoch time at which it ended the model training.

- **run time**: This value indicates the epoch total time it took to finish training the model.

- **default_threshold**: This is the default threshold value over which the model will consider all predictions as **1**. Similarly, any value predicted below the default threshold will be **0**.

The following screenshot shows you the output sub-section of the model details:

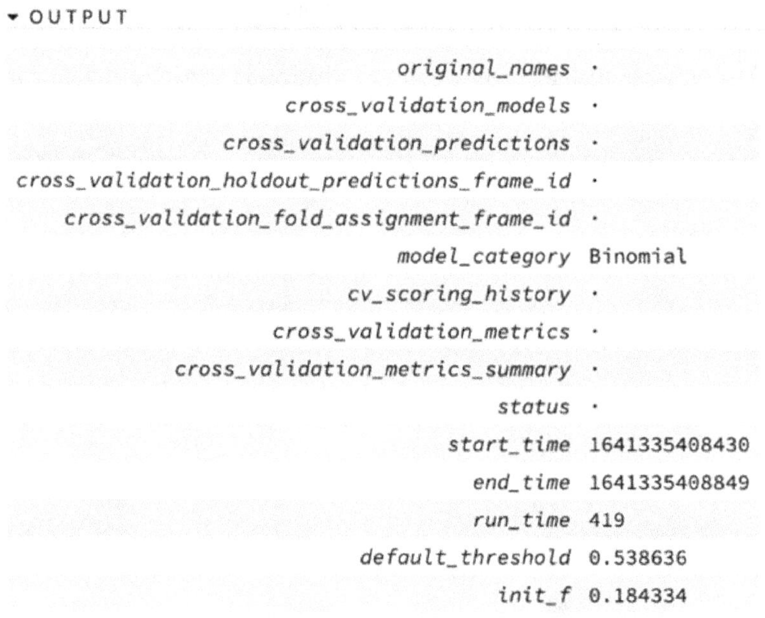

```
▼ OUTPUT

                                   original_names  ·
                            cross_validation_models  ·
                       cross_validation_predictions  ·
     cross_validation_holdout_predictions_frame_id  ·
        cross_validation_fold_assignment_frame_id  ·
                                    model_category  Binomial
                                 cv_scoring_history  ·
                           cross_validation_metrics  ·
                   cross_validation_metrics_summary  ·
                                            status  ·
                                        start_time  1641335408430
                                          end_time  1641335408849
                                          run_time  419
                                 default_threshold  0.538636
                                            init_f  0.184334
```

Figure 2.36 – Output of the model information

- **Column_Types**: This sub-section gives you a brief look at the column types of the columns in the dataframe. The values are in the same sequence as the sequence of the columns in the training dataframe.

The following screenshot shows you the **column_types** metrics of the model details:

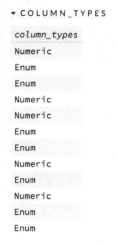

▾ COLUMN_TYPES

column_types
Numeric
Enum
Enum
Numeric
Numeric
Enum
Enum
Numeric
Enum
Numeric
Enum
Enum

Figure 2.37 – Column types

- **Output - Training_Metrics**: This sub-section shows you the metrics of the model when the training dataset is used for predictions. We will learn about the various metrics of ML in future chapters.

The following screenshot shows you the training metrics of the model details:

▾ OUTPUT · TRAINING_METRICS

model	GBM_1_AutoML_1_20220104_223006
model_checksum	1667669939565867160
frame	AutoML_1_20220104_223006_training_training_dataset
frame_checksum	480731267713440174
description	·
model_category	Binomial
scoring_time	1641335408839
predictions	·
MSE	0.086861
RMSE	0.294722
nobs	729
custom_metric_name	·
custom_metric_value	0
r2	0.649596
logloss	0.294230
AUC	0.947350
pr_auc	0.949081
Gini	0.894700
mean_per_class_error	0.108940

Figure 2.38 – The model output's training metrics

- **Output - Validation_Metrics**: Similar to the previous sub-section, this sub-section shows you the metrics of the model when the validation dataset is used for predictions.

The following screenshot shows you the training metrics of the model details:

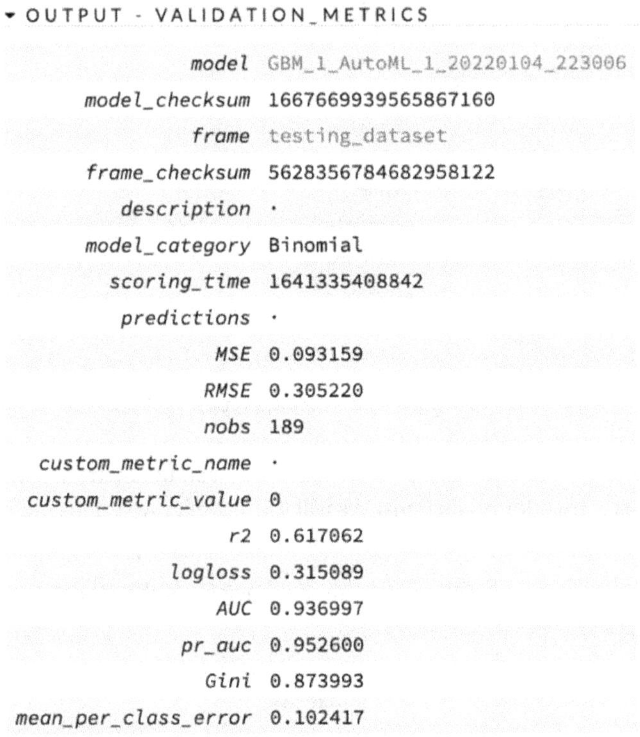

```
▼ OUTPUT - VALIDATION_METRICS

                  model  GBM_1_AutoML_1_20220104_223006
         model_checksum  16676699939565867160
                  frame  testing_dataset
         frame_checksum  5628356784682958122
            description  ·
         model_category  Binomial
           scoring_time  1641335408842
            predictions  ·
                    MSE  0.093159
                   RMSE  0.305220
                   nobs  189
      custom_metric_name  ·
     custom_metric_value  0
                     r2  0.617062
                logloss  0.315089
                    AUC  0.936997
                 pr_auc  0.952600
                   Gini  0.873993
   mean_per_class_error  0.102417
```

Figure 2.39 – The model output's validation metrics

Now that we have understood the various parts of the model details, let's see how we can perform predictions on this newly trained model.

Working with prediction functions in H2O Flow

Now that you finally have a trained model, we can perform predictions on it. Predictions on trained models are straightforward. You just need to load the model and pass in your dataset, which contains the data on which you want to make predictions. H2O will use the loaded model and make predictions for all the values in the dataset. Let's use the `prediction_dataframe.hex` dataframe that we created previously to make predictions on.

We will gain an understanding of the prediction operations in the following sub-sections, starting with gaining an understanding of how to make predictions.

Making predictions using H2O Flow

First, let's start by exploring the **Score** operation's drop-down list in the topmost part of the web UI.

You will see a list of scoring operations, as follows:

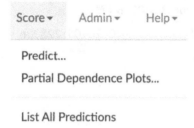

Figure 2.40 – The Score functions drop-down menu

The preceding drop-down menu shows you a list of all the various scoring operations that you can perform.

The functions are listed as follows:

- **Predict**: This makes predictions using trained models.
- **Partial Dependency Plots**: These show you a graphical representation of the effects of a variable on the predictions. In other words, they show you how changing certain variables in the data used for prediction affects the prediction response.
- **List all Predictions**: This function lists all the predictions that were made.

There are two ways you can start making predictions in H2O Flow. They are listed as follows:

- Use the **Predict** function from the **Score** operation's drop-down menu.
- Click on the **Predict** button in the **Actions** section of a model, as shown in *Figure 2.34*.

Both methods will give the same output, as shown in the following screenshot:

```
predict model: "GBM_1_AutoML_1_20220104_223006"
```

⚡ Predict

Name:	prediction-8c7f82d6-f7
Model:	GBM_1_AutoML_1_20220104_223006
Frame:	testing_dataset ⌄
Compute Leaf Node Assignment:	☐
Actions:	⚡ Predict

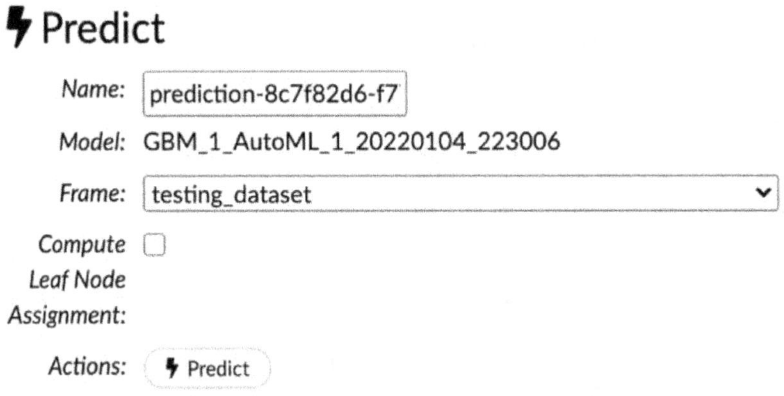

Figure 2.41 – Prediction

Clicking on the ML model's **Predict** action button will set the **Model** parameter to the respective model ID. However, when clicking on the **Predict** operation button in the **Scores** operation drop-down list, you get the option to select any model.

The **Predict** operation will prompt you for the following parameters:

- **Name**: You can name the prediction result for easier identification. This is especially handy if you are experimenting with different prediction requests and need a quick referral to a specific prediction result of your interest.

- **Model**: This is the ID of the model you wish to use to make the prediction. Since we selected the **Predict** action of our **Gradient Boosting Machine** (**GBM**) model, this will be non-editable.

- **Frame**: This is the dataframe that you want to use to make predictions on. So, let's select the **prediction_dataset.hex** file.

- **Compute Leaf Node Assignment**: This returns the leaf placements of the row in all the trees of the model for every row in the dataframe used to make the prediction. For our walk-through, we won't need this, and as such, we can leave it unchecked.

  Once you have selected the appropriate parameter values, the only thing left to do is to click on the **Predict** button to make your predictions.

  The following screenshot shows you the output of the **Predict** operation:

Figure 2.42 – The Prediction output

Congratulations! You have finally managed to make a prediction on a model trained by AutoML. Let's move on to the next section, where we will explore and understand the prediction results that we just got.

Understanding the prediction results

Prediction is the final stage of the ML pipeline. It is the part that brings the actual value to all the efforts put into creating the ML pipeline. Making predictions is easy; however, it is important to understand what the actual predicted value is and what it represents against the input values.

The prediction result gives you detailed information on not only the predicted values but also metric information and certain metadata, as shown in *Figure 2.42*.

The following screenshot shows you the prediction output sub-section of your prediction results:

Figure 2.43 – The prediction results

This is similar to the **output - validation and training metrics** of the model details that we saw in *Figure 2.38* and *Figure 2.39*.

Inside this sub-section, you have the prediction frame link next to the predictions key. Clicking on it shows you the frame summary of your prediction values. Let's do that so that we can get a good look at the prediction values.

The following screenshot shows you the prediction summary details in the form of a dataframe:

Figure 2.44 – The prediction dataframe

H2O stores prediction results as dataframes, too. Therefore, they have the same features that we discussed in the *Working with Data Functions in H2O Flow* section.

The **COLUMN SUMMARIES** section indicates the three columns in the prediction dataframe. They are listed as follows:

- **Predict**: This column is of the enum type and indicates the prediction value for the input rows in the prediction dataframe.

- **P0**: This column indicates the probability that the predicted value is 0.

- **P1**: This column indicates the probability that the predicted value is 1.

Let's view the data to better understand its contents. You can view the content of the dataframe by clicking on the **View Data** button in the **Actions** section of the prediction dataframe output, as shown in *Figure 2.44*.

The following screenshot shows you the expected output when viewing the prediction data:

Figure 2.45 – The content of the prediction dataframe

Let's return to the **Prediction** sub-section of the prediction output, as shown in *Figure 2.43*. Let's combine the prediction results with the dataframe we used as input for the predictions. You can do this by clicking on the **Combine predictions with frame** button.

The following screenshot shows you the output of that operation:

Figure 2.46 – The Combine predictions with frame result

Clicking on the **View Frame** button shows you the combined frame, which is also a dataframe. Let's click on the **View Frame** button to view the contents of the frame.

The following screenshot shows you the output of the **View Frame** operation from the frames' combined output:

Figure 2.47 – The dataframe used for prediction combined with prediction results

Selecting the **View data** button in the **Actions** section of the combined prediction output shows you the complete contents of the dataframe with the predicted values next to it.

The following screenshot shows you the contents of the combined dataframes:

```
cs   getFrameData "combined-prediction-8c7f82d6-f774-4c9a-9e25-37d408edd167"
```

⊞ combined-prediction-8c7f82d6-f774-4c9a-9e25-37d408edd167

▾ DATA

Row	predict	p0	p1	Age	Sex	ChestPoinType	RestingBP	Cholesterol	FastingBS	RestingECG	MaxHR	ExerciseAngina	Oidpeak	ST_Slope	HeartDisease
1	0	0.9138	0.0862	39.0	M	NAP	120.0	339.0	0	Normal	170.0	N	0	Up	0
2	1	0.3062	0.6938	38.0	M	ASY	110.0	196.0	0	Normal	166.0	N	0	Flat	1
3	1	0.1320	0.8680	41.0	M	ASY	130.0	172.0	0	ST	130.0	N	2.0	Flat	1
4	0	0.9404	0.0596	32.0	M	ATA	125.0	254.0	0	Normal	155.0	N	0	Up	0
5	0	0.9589	0.0411	50.0	M	ATA	140.0	216.0	0	Normal	170.0	N	0	Up	0
6	1	0.2106	0.7894	47.0	F	ASY	120.0	205.0	0	Normal	98.0	Y	2.0	Flat	1
7	1	0.1018	0.8982	52.0	M	ASY	112.0	342.0	0	ST	96.0	Y	1.0	Flat	1
8	0	0.9369	0.0631	49.0	M	ATA	100.0	253.0	0	Normal	174.0	N	0	Up	0
9	1	0.0344	0.9656	52.0	M	ASY	160.0	246.0	0	ST	82.0	Y	4.0	Flat	1
10	0	0.7541	0.2459	44.0	M	ASY	150.0	412.0	0	Normal	170.0	N	0	Up	0
11	1	0.2028	0.7972	63.0	M	ASY	150.0	223.0	0	Normal	115.0	N	0	Flat	1
12	0	0.9469	0.0531	52.0	M	ATA	160.0	196.0	0	Normal	165.0	N	0	Up	0
13	1	0.4461	0.5539	43.0	M	TA	120.0	291.0	0	ST	155.0	N	0	Flat	1
14	0	0.8049	0.1951	39.0	M	ASY	130.0	307.0	0	Normal	140.0	N	0	Up	0
15	1	0.2738	0.7262	46.0	M	ASY	118.0	186.0	0	Normal	124.0	N	0	Flat	1
16	0	0.7119	0.2881	50.0	M	ASY	140.0	129.0	0	Normal	135.0	N	0	Up	0
17	1	0.0266	0.9734	57.0	M	ASY	150.0	255.0	0	Normal	92.0	Y	3.0	Flat	1
18	1	0.3805	0.6195	33.0	F	ASY	100.0	246.0	0	Normal	150.0	N	1.0	Flat	1
19	1	0.1011	0.8989	59.0	F	ASY	130.0	338.0	1	ST	130.0	Y	1.5000	Flat	1
20	0	0.9710	0.0290	34.0	F	ATA	130.0	161.0	0	Normal	190.0	N	0	Up	0
21	0	0.7275	0.2725	48.0	F	ASY	108.0	163.0	0	Normal	175.0	N	2.0	Up	0
22	1	0.0254	0.9746	56.0	M	ASY	170.0	388.0	0	ST	122.0	Y	2.0	Flat	1
23	0	0.7764	0.2236	39.0	M	ASY	110.0	273.0	0	Normal	132.0	N	0	Up	0

Figure 2.48 – The content of the combined dataframes

Observing the contents of the combined dataframe, you will notice that you can now easily compare the predicted values, as mentioned in the **predict** column, and compare them with the actual value in the **HeartDisease** column on the same row.

Congratulations! You have officially made predictions on a model that you trained and combined the results into a single dataframe for a comparative view, which you can share with your stakeholders. So, this sums up our walk-through of how to create an ML pipeline using H2O Flow.

Summary

In this chapter, we understood the various functionality that H2O Flow has to offer. After getting comfortable with the web UI, we started implementing our ML pipeline. We imported and parsed the Heart Failure Prediction dataset. We understood the various operations that can be performed on the dataframe, understood the metadata and statistics of the dataframe, and prepared the dataset to later train, validate, and predict models.

Then, we trained models on the dataframe using AutoML. We understood the various parameters that needed to be input to correctly configure AutoML. We trained models using AutoML and understood the leaderboard. Then, we dived deeper into the details of the models trained and tried our best to understand their characteristics.

Once our model was trained, we performed predictions on it and then explored the prediction output by combining it with the original dataframe so that we could compare the predicted values.

In the next chapter, we will explore the various data manipulation operations further. This will help us to gain an understanding of what steps need to be taken in terms of cleaning and transforming the dataframe and how it could improve the quality of the models trained.

Part 2
H2O AutoML
Deep Dive

This part will help you understand the inner workings of H2O AutoML. This will involve how H2O AutoML handles data processing, training, and the selection of models, and how it measures the performance of trained models. This part will also help you understand how to read the various performance graphs and other model details that will help make sense of the models' behavior. All of this will help you further experiment and explore H2O AutoML and get the most out of it based on your needs.

This section comprises the following chapters:

- *Chapter 3, Understanding Data Processing*
- *Chapter 4, Understanding H2O AutoML Training and Architecture*
- *Chapter 5, Understanding AutoML Algorithms*
- *Chapter 6, Understanding H2O AutoML Leaderboard and Other Performance Metrics*
- *Chapter 7, Working with Model Explainability*

3
Understanding Data Processing

A **Machine Learning** (ML) model is the output we get once data is fitted into an ML algorithm. It represents the underlying relationship between various features and how that relationship impacts the target variable. This relationship depends entirely on the contents of the dataset. What makes every ML model unique, despite using the same ML algorithm, is the dataset that is used to train said model. Data can be collected from various sources and can have different schemas and structures, which need not be structurally compatible among themselves but may in fact be related to each other. This relationship can be very valuable and can also potentially be the differentiator between a good and a bad model. Thus, it is important to transform this data to meet the requirements of the ML algorithm to eventually train a good model.

Data processing, data preparation, and data preprocessing are all steps in the ML pipeline that focus on best exposing the underlying relationship between the features by transforming the structure of the data. Data processing may be the most challenging step in the ML pipeline, as there are no set steps to the transformation process. Data processing depends entirely on the problem you wish to solve; however, there are some similarities among all datasets that can help us define certain processes that we can perform to optimize our ML pipeline.

In this chapter, we will learn about some of the common functionalities that are often used in data processing and how H2O has in-built operations that can help us easily perform them. We will understand some of the H2O operations that can reframe the structure of our dataframe. We will understand how to handle missing values and the importance of the imputation of values. We will then investigate how we can manipulate the various feature columns in the dataframe, as well as how to slice the dataframe for different needs. We shall also investigate what encoding is and what the different types of encoding are.

In this chapter, we are going to cover the following main topics:

- Reframing your dataframe
- Handling missing values in the dataframe
- Manipulation of feature columns of the dataframe

- Tokenization of textual data
- Encoding of data using target encoding

Technical requirements

All code examples in this chapter are run on **Jupyter Notebook** for an easy understanding of what each line in the code block does. You can run the whole block of code via a Python or R script executor and observe the output results, or you can follow along by installing Jupyter Notebook and observing the execution results of every line in the code blocks.

To install Jupyter Notebook, make sure you have the latest version of Python and `pip` installed on your system and execute the following command:

```
pip install jupyterlab
```

Once JupyterLab has successfully installed, you can start your Jupyter Notebook locally by executing the following command in your terminal:

```
jupyter notebook
```

This will open the **Jupyter Notebook** page on your default browser. You can then select which language you want to use and start executing the lines in the code step by step.

All code examples for this chapter can be found on GitHub at `https://github.com/PacktPublishing/Practical-Automated-Machine-Learning-on-H2O/tree/main/Chapter%203`.

Now, let's begin processing our data by first creating a dataframe and reframing it so that it meets our model training requirement.

Reframing your dataframe

Data collected from various sources is often termed **raw data**. It is called raw in the sense that there might be a lot of unnecessary or stale data, which might not necessarily benefit our model training. The structure of the data collected also might not be consistent among all the sources. Hence, it becomes very important to first reframe the data from various sources into a consistent format.

You may have noticed that once we import the dataset into H2O, H2O converts the dataset into a `.hex` file, also called a dataframe. You have the option to import multiple datasets as well. Assuming you are importing multiple datasets from various sources, each with its own format and structure, then you will need a certain functionality that helps you reframe the contents of the dataset and merge them to form a single dataframe that you can feed to your ML pipeline.

H2O provides several functionalities that you can use to perform the required manipulations.

Here are some of the dataframe manipulation functionalities that help you reframe your dataframe:

- Combining columns from two dataframes
- Combining rows from two dataframes
- Merging two dataframes

Let's see how we can combine columns from different dataframes in H2O.

Combining columns from two dataframes

One of the most common dataframe manipulation functionalities is combining different columns from different dataframes. Sometimes, the columns of one dataframe may be related to those of another. This could prove beneficial during model training. Thus, it is quite useful to have a functionality that can help us manipulate these columns and combine them together to form a single dataframe for model training.

H2O has a function called cbind() that combines the columns from one dataset into another.

Let's try this function out in our Jupyter Notebook using Python. Execute the following steps in sequence:

1. Import the h2o library:

    ```
    import h2o
    ```

2. Import the numpy library; we will use this to create a sample dataframe for our study:

    ```
    import numpy as np
    ```

3. Initialize the h2o server:

    ```
    h2o.init()
    ```

4. Now, let's create a dataframe called important_dataframe_1; this is a dataframe whose columns are important. To ensure that you generate the same values in the dataset as in this example, set the random seed value for numpy to 123. We will set the number of rows to 15 and the number of columns to 5. You can name the columns anything you like:

    ```
    np.random.seed(123)
    important_dataframe_1 = h2o.H2OFrame.from_python(np.
    random.randn(15,5).tolist(), column_names=list(["
    important_column_1" , " important_column_2" , "
    important_column_3" , " important_column_4" , "
    important_column_5" ]))
    ```

5. Let's check out the content of the dataset by executing the following code:

```
important_dataframe_1.describe
```

The following screenshot shows you the contents of the dataset:

important_column_1	important_column_2	important_column_3	important_column_4	important_column_5
-1.08563	0.997345	0.282978	-1.50629	-0.5786
1.65144	-2.42668	-0.428913	1.26594	-0.86674
-0.678886	-0.094709	1.49139	-0.638902	-0.443982
-0.434351	2.20593	2.18679	1.00405	0.386186
0.737369	1.49073	-0.935834	1.17583	-1.25388
-0.637752	0.907105	-1.42868	-0.140069	-0.861755
-0.255619	-2.79859	-1.77153	-0.699877	0.927462
-0.173636	0.00284592	0.688223	-0.879536	0.283627
-0.805367	-1.72767	-0.3909	0.573806	0.338589
-0.0118305	2.39237	0.412912	0.978736	2.23814

Figure 3.1 – important_dataframe_1 data content

6. Let's create another dataframe called important_dataframe_2, as before but with different column names, but an equal number of rows and only 2 columns:

```
important_dataframe_2 = h2o.H2OFrame.from_python(np.
random.randn(15,2).tolist(), column_names=list(["
important_column_6" , " important_column_7" ]))
```

7. Let's check out the content of this dataframe as well:

important_column_6	important_column_7
1.03973	-0.403366
-0.12603	-0.837517
-1.60596	1.25524
-0.688869	1.66095
0.807308	-0.314758
-1.0859	-0.732462
-1.21252	2.08711
0.164441	1.15021
-1.26735	0.181035
1.17786	-0.335011

Figure 3.2 – important_dataframe_2 data content

8. Now, let's combine the columns of both the dataframes and store them in another variable called `final_dataframe`, using the `cbind()` function:

```
final_dataframe = important_dataframe_1.cbind(important_
dataframe_2)
```

9. Let's now observe `final_dataframe`:

```
final_dataframe.describe
```

You should see the contents of **final_dataframe** as follows:

important_column_1	important_column_2	important_column_3	important_column_4	important_column_5	important_column_6	important_column_7
-1.08563	0.997345	0.282976	-1.50629	-0.5786	1.03973	-0.403366
1.65144	-2.42668	-0.428913	1.26594	-0.86674	-0.12603	-0.837517
-0.678886	-0.094709	1.49139	-0.638902	-0.443982	-1.60596	1.25524
-0.434351	2.20593	2.18679	1.00405	0.386186	-0.688869	1.66095
0.737369	1.49073	-0.935834	1.17583	-1.25388	0.807308	-0.314758
-0.637752	0.907105	-1.42868	-0.140069	-0.861755	-1.0859	-0.732462
-0.255619	-2.79859	-1.77153	-0.699877	0.927462	-1.21252	2.08711
-0.173636	0.00284592	0.688223	-0.879536	0.283627	0.164441	1.15021
-0.805367	-1.72767	-0.3909	0.573806	0.338589	-1.26735	0.181035
-0.0118305	2.39237	0.412912	0.978736	2.23814	1.17786	-0.335011

Figure 3.3 – final_dataframe data content after cbind()

Here, you will notice that we have successfully combined the columns from `important_dataframe_2` with the columns of **important_dataframe_1**.

This is how you can use the `cbind()` function to combine the columns of two different datasets into a single dataframe. The only thing to bear in mind while using the `cbind()` function is that it is necessary to ensure that both the datasets to be combined have the same number of rows. Also, if you have dataframes with the same column name, then H2O will append a **0** in front of the column from dataframe.

Now that we know how to combine the columns of different dataframes, let's see how we can combine the column values of multiple dataframes with the same column structure.

Combining rows from two dataframes

The majority of big corporations often handle tremendous amounts of data. This data is often partitioned into multiple chunks to make storing and reading it faster and more efficient. However, during model training, we will often need to access all these partitioned datasets. These datasets have the same structure but the data contents are distributed. In other words, the dataframes have the same columns; however, the data values or rows are split among them. We will often need a function that combines all these dataframes together so that we have all the data values available for model training.

H2O has a function called `rbind()` that combines the rows from one dataset into another.

Let's try this function out in the following example:

1. Import the h2o library:

```
import h2o
```

2. Import the numpy library; we will use this to create a random dataframe for our study:

```
import numpy as np
```

3. Initialize the h2o server:

```
h2o.init()
```

4. Now, let's create a random dataframe called `important_dataframe_1`. To ensure that you generate the same values in the dataset as in this example, set the random seed value for numpy to `123`. We will set the number of rows to `15` and the number of columns to `5`. You can name the columns anything you like:

```
np.random.seed(123)
important_dataframe_1 = h2o.H2OFrame.from_python(np.
random.randn(15,5).tolist(), column_names=list(["
important_column_1" , " important_column_2" ," important_
column_3" ," important_column_4" ," important_column_5"
]))
```

5. Let's check out the number of rows of the dataframe, which should be `15`:

```
important_dataframe_1.nrows
```

6. Let's create another dataframe called `important_dataframe_2`, as with the previous one, with the same column names and any number of rows. In the example, I have used `10` rows:

```
important_dataframe_2 = h2o.H2OFrame.from_python(np.
random.randn(10,5).tolist(), column_names=list(["
important_column_1" , " important_column_2" ," important_
column_3" ," important_column_4" ," important_column_5"
]))
```

7. Let's check out the number of rows for `important_dataframe_2`, which should be `10`:

```
important_dataframe_2.nrows
```

8. Now, let's combine the rows of both the dataframes and store them in another variable called `final_dataframe`, using the `rbind()` function:

```
final_dataframe = important_dataframe_1.rbind(important_
dataframe_2)
```

9. Let's now observe `final_dataframe`:

```
final_dataframe.describe
```

You should see the contents of **final_dataframe** as follows:

important_column_1	important_column_2	important_column_3	important_column_4	important_column_5
-1.08563	0.997345	0.282978	-1.50629	-0.5786
1.65144	-2.42668	-0.428913	1.26594	-0.86674
-0.678886	-0.094709	1.49139	-0.638902	-0.443982
-0.434351	2.20593	2.18679	1.00405	0.386186
0.737369	1.49073	-0.935834	1.17583	-1.25388
-0.637752	0.907105	-1.42868	-0.140069	-0.861755
-0.255619	-2.79859	-1.77153	-0.699877	0.927462
-0.173636	0.00284592	0.688223	-0.879536	0.283627
-0.805367	-1.72767	-0.3909	0.573806	0.338589
-0.0118305	2.39237	0.412912	0.978736	2.23814

Figure 3.4 – final_dataframe data contents after rbind()

10. Let's check out the number of rows in **final_dataframe**:

```
final_dataframe.nrows
```

The output of the last operation should show you the value of the number of rows in the final dataset. You will see that the value is **25** and the contents of the dataframe are the combined row values of both the previous datasets.

Now that we have understood how to combine the rows of two dataframes in H2O using the `rbind()` function, let's see how we can fully combine two datasets.

Merging two dataframes

You can directly merge two dataframes, combining their rows and columns into a single dataframe. H2O provides a merge() function that combines two datasets that share a common column or common columns. During merging, columns that the two datasets have in common are used as the **merge key**. If they only have one column in common, then that column forms the singular primary key for the merge. If there are multiple common columns, then H2O will form a complex key of all these columns based on their data values and use that as the merge key. If there are multiple common columns between the two datasets and you only wish to merge a specific subset of them, then you will need to rename the other common columns to remove the corresponding commonality.

Let's try this function out in the following example in Python:

1. Import the h2o library:

    ```
    import h2o
    ```

2. Import the numpy library; we will use this to create a random dataframe for our study:

    ```
    import numpy as np
    ```

3. Initialize the h2o server:

    ```
    h2o.init()
    ```

4. Now, let's create a dataframe called dataframe_1. The dataframe has 3 columns: words, numerical_representation, and letters. Now, let's fill in the data content as follows:

    ```
    dataframe_1 = h2o.H2OFrame.from_python({'words':['Hello',
    'World', 'Welcome', 'To', 'Machine', 'Learning'],
    'numerical_representation': [0,1,2,3,4,5],'letters':['a',
    'b','c','d']})
    ```

5. Let's check out the content of the dataset:

    ```
    dataframe_1.describe
    ```

6. You will notice the contents of the dataset as follows:

words	numerical_representation	letters
Hello	0	a
World	1	b
Welcome	2	c
To	3	d
Machine	4	
Learning	5	

Figure 3.5 – dataframe_1 data content

7. Let's create another dataframe called dataframe_2. This dataframe also contains 3 columns: the numerical_representation column, the letters column (both of which it has in common with dataframe_1), and an uncommon column. Let's call it other_words:

```
dataframe_2 = h2o.H2OFrame.from_python({'other_
words':['How', 'Are', 'You', 'Doing', 'Today',
'My', 'Friend', 'Learning', 'H2O', 'Artificial',
'Intelligence'], 'numerical_representation':
[0,1,2,3,4,5,6,7,8,9],'letters':['a','b','c','d','e']})
```

8. Let's check out the content of this dataframe as well:

```
dataframe_2.head(11)
```

On executing the code, you should see the following output in your notebook:

other_words	numerical_representation	letters
How	0	a
Are	1	b
You	2	c
Doing	3	d
Today	4	e
My	5	
Friend	6	
Learning	7	
H2O	8	
Artificial	9	
Intelligence	nan	

Figure 3.6 – dataframe_2 data contents

9. Now, let's merge `dataframe_1` into `dataframe_2`, using the `merge()` operation:

    ```
    final_dataframe = dataframe_2.merge(dataframe_1)
    ```

10. Let's now observe `final_dataframe`:

    ```
    final_dataframe.describe
    ```

11. You should see the contents of **final_dataframe** as follows:

numerical_representation	letters	other_words	words
0	a	How	Hello
1	b	Are	World
2	c	You	Welcome
3	d	Doing	To
5		My	Learning

Figure 3.7 – final_dataframe contents after merge()

You will notice that H2O used the combination of the **numerical_representation** column and the **letters** column as the merging key. This is why we have values ranging from **1 to 5** in the `numerical_representation` column with the appropriate values in the other columns.

Now, you may be wondering why there is no row for **4**. That is because while merging, we have two common columns: **numerical_representation** and **letters**. So, H2O used a complex merging key that uses both these columns: **(0, a)**, **(1, b)**, **(2, c)**, and so on.

Now the next question you might have is *What about the row with the value 5? It has no value in the letters column*. That is because even an empty value is treated as a unique value in ML. Thus, during merging, the complex key that was generated treated **(5,)** as a valid merge key.

H2O drops all the remaining values since **dataframe_1** does not have any more numerical representation values.

12. You can enforce H2O to not drop any of the values from the merge key column by setting the `all_x` parameter to `True` as follows:

    ```
    final_dataframe = dataframe_2.merge(dataframe_1, all_x =
    True)
    ```

13. Now, let's observe the contents of **final_dataframe** by using its describe attribute:

numerical_representation	letters	other_words	words
0	a	How	Hello
1	b	Are	World
2	c	You	Welcome
3	d	Doing	To
4	e	Today	
5		My	Learning
6		Friend	
7		Learning	
8		Artificial	
9		Intelligence	

Figure 3.8 – final_dataframe data content after enforcing merge()

You will notice that we now have all the values from both dataframes merged into a single dataframe. We have all the numerical representations from **0 to 9** and all letters from **a to e** from **dataframe_2** that were missing in the previous step, along with the correct values from the **other_words** column and the **words** column.

To recap, we learned how to combine dataframe columns and rows. We also learned how to combine entire dataframes together using the merge() function. However, we noticed that if we enforced the merging of dataframes despite them not having common data values in their key columns, we ended up with missing values in the dataframe.

Now, let's look at the different methods we can use to handle missing values using H2O.

Handling missing values in the dataframe

Missing values in datasets are the most common issue in the real world. It is often expected to have at least a few instances of missing data in huge chunks of datasets collected from various sources. Data can be missing for several reasons, which can range from anything from data not being generated at the source all the way to downtimes in data collectors. Handling missing data is very important for model training, as many ML algorithms don't support missing data. Those that do may end up giving more importance to looking for patterns in the missing data, rather than the actual data that is present, which distracts the machine from learning.

Missing data is often referred to as **Not Available** (**NA**) or **nan**. Before we can send a dataframe for model training, we need to handle these types of values first. You can either drop the entire row that contains any missing values or you can fill them with any default value either default or common for that data column. How you handle missing values depends entirely on which data is missing and how important it is for the overall model training.

H2O provides some functionalities that you can use to handle missing values in a dataframe. These are some of them:

- The `fillna()` function
- Replacing values in a frame
- Imputation

Next, let's see how we can fill missing values in a dataframe using H2O.

Filling NA values

`fillna()` is a function in H2O that you can use to fill missing data values in a sequential manner. This is especially handy if you have certain data values in a column that are sequential in nature, for example, time series or any metric that increases or decreases sequentially and can be sorted. The smaller the difference between the values in the sequence, the more applicable this function becomes.

The `fillna()` function has the following parameters:

- `method`: This can either be *forward* or *backward*. It indicates the direction in which H2O should start filling the NA values in the dataframe.
- `axis`: 0 for column-wise fill or 1 for row-wise fill.
- `maxlen`: The maximum number of consecutive NAs to fill.

Let's see an example in Python of how we can use this function to fill missing values:

1. Import the h2o library:

   ```
   import h2o
   ```

2. Import the numpy library; we will use this to create a random dataframe for our study:

   ```
   import numpy as np
   ```

3. Initialize the h2o server:

   ```
   h2o.init()
   ```

4. Create a random dataframe with 1000 rows, 3 columns, and some NA values:

```
dataframe = h2o.create_frame(rows=1000, cols=3, integer_
fraction=1.0, integer_range=100, missing_fraction=0.2,
seed=123)
```

5. Let's observe the contents of this dataframe. Execute the following code and you will see certain missing values in the dataframe:

```
dataframe.describe
```

You should see the contents of the dataframe as follows:

C1	C2	C3
nan	nan	77
94	14	-58
94	-26	-39
96	93	-56
-85	44	-53
58	-28	27
-84	63	92
78	-65	94
nan	81	nan
-27	-61	13

Figure 3.9 – Dataframe contents

6. Let's now use the `fillna()` function to forward fill the NA values. Execute the following code:

```
filled_dataframe = dataframe.fillna(method=" forward" ,
axis=0, maxlen=1)
```

7. Let's observe the filled contents of the dataframe. Execute the following code:

```
filled_dataframe.describe
```

8.   You should see the contents of the dataframe as follows:

C1	C2	C3
nan	nan	77
94	14	-58
94	-26	-39
96	93	-56
-85	44	-53
58	-28	27
-84	63	92
78	-65	94
78	81	94
-27	-61	13

Figure 3.10 – filled_dataframe contents

The `fillna()` function has filled most of the NA values in the dataframe sequentially.

However, you will notice that we still have some **nan** values in the first row of the dataframe. This is because we filled the dataframe missing values row-wise in the **forward** direction. When filling NA values, H2O will record the last value in a row for a specific column and copy it if the value in the subsequent row is NA. Since this is the very first column, H2O does not have any previous value in the record to fill it, thus it skips over it.

Now that we understand how we can sequentially fill data in a dataframe using the `fillna()` function in H2O, let's see how we can replace certain values in the dataframe.

Replacing values in a frame

Another common functionality often needed for data processing is replacing certain values in the dataframe. There can be plenty of reasons why you might want to do this. This is especially common for numerical data where some of the most common transformations include rounding off values, normalizing numerical ranges, or just correcting a data value. In this section, we will explore some of the functions that we can use in H2O to replace values in the dataframe.

Let's first create a dataframe that we can use to test out such functions. Execute the following code so that we have a dataframe ready for manipulation:

```
import h2o
h2o.init()
dataframe = h2o.create_frame(rows=10, cols=3, real_range=100,
```

```
integer_fraction=1, missing_fraction=0.1, seed=5)
dataframe.describe
```

The dataframe should look as follows:

C1	C2	C3
-99	-18	75
-73	-10	58
-33	nan	70
21	nan	-61
-52	20	79
-22	-98	19
nan	-58	52
-58	-31	-9
79	-26	nan
12	-81	66

Figure 3.11 – Dataframe data contents

So, we have a dataframe with three columns: **C1**, **C2**, and **C3**. Each column has a few negative numbers and some **nan** values. Let's see how we can play around with this dataframe.

Let's start with something simple. Let's update the value of a single data value, also called a **datum**, in the dataframe. Let's update the fourth row of the **C2** column to 99. You can update the value of a single data value based on its position in the dataframe as follows:

```
dataframe[3,1] = 99
```

Note that the columns and rows in the dataframe all start with 0. Hence, we set the value in the dataframe with the row number of 3 and the column number of 1 as 99. You should see the results in the dataframe by executing dataframe.describe as follows:

```
dataframe.describe
```

The dataframe should look as follows:

C1	C2	C3
-99	-18	75
-73	-10	58
-33	nan	70
21	99	-61
-52	20	79
-22	-98	19
nan	-58	52
-58	-31	-9
79	-26	nan
12	-81	66

Figure 3.12 – Dataframe contents after the datum update

As you can see in the dataframe, we replaced the **nan** value that was previously in the third row of the **C2** column with **99**.

This is a manipulation of just one data value. Let's see how we can replace the values of an entire column. Let's increase the data values in the **C3** column to three times their original value. You can do so by executing the following code:

```
dataframe[2] = 3*dataframe[2]
```

You should see the results in the dataframe by executing `dataframe.describe` as follows:

```
dataframe.describe
```

The dataframe should look as follows:

C1	C2	C3
-99	-18	225
-73	-10	174
-33	nan	210
21	99	-183
-52	20	237
-22	-98	57
nan	-58	156
-58	-31	-27
79	-26	nan
12	-81	198

Figure 3.13 – Dataframe contents after column value updates

We can see in the output that the values in the **C3** column have now been increased to three times the original values in the column.

All these replacements we performed till now are straightforward. Let's try some conditional updates on the dataframe. Let's round off all the negative numbers in the dataframe to 0. So, the condition is that we only update the negative numbers to 0 and don't change any of the positive numbers. You can do conditional updates as follows:

```
dataframe[dataframe['C1'] < 0, "C1"] = 0
dataframe[dataframe['C2'] < 0, "C2"] = 0
dataframe[dataframe['C2'] < 0, "C3"] = 0
```

You should see the results in the dataframe by executing `dataframe.describe` as follows:

```
dataframe.describe
```

The dataframe should look as follows:

C1	C2	C3
0	0	225
0	0	174
0	nan	210
21	99	-183
0	20	237
0	0	57
nan	0	156
0	0	-27
79	0	nan
12	0	198

Figure 3.14 – Dataframe contents after conditional updates

As you can see in the dataframe, all the negative values have been rounded up/replaced by **0**.

Now, what if instead of rounding the negative numbers up to **0** we wished to just inverse the negative numbers? We could do so by combining the conditional updates with arithmetic updates. Refer to the following example:

```
dataframe[" C1" ] = (dataframe[" C1" ] < 0).ifelse(-
1*dataframe[" C1" ], dataframe[" C1" ])
dataframe[" C2" ] = (dataframe[" C2" ] < 0).ifelse(-
1*dataframe[" C2" ], dataframe[" C2" ])
dataframe[" C3" ] = (dataframe[" C3" ] < 0).ifelse(-
1*dataframe[" C3" ], dataframe[" C3" ])
```

Now, let's try to see whether we can replace the remaining **nan** values with something valid. We already read about the `fillna()` function, but what if the **nan** values are nothing but some missing values that don't exactly fall into any incremental or decremental pattern, and we just want to set it to 0? Let's do that now. Run the following code:

```
dataframe[dataframe[" C1" ].isna(),  " C1" ] = 0
dataframe[dataframe[" C2" ].isna(),  " C2" ] = 0
dataframe[dataframe[" C3" ].isna(),  " C3" ] = 0
```

You should see the results in the dataframe by executing `dataframe.describe` as follows:

```
dataframe.describe
```

The dataframe should look as follows:

C1	C2	C3
0	0	225
0	0	174
0	0	210
21	99	183
0	20	237
0	0	57
0	0	156
0	0	27
79	0	0
12	0	198

Figure 3.15 – Dataframe contents after replacing nan values with 0

The `isna()` function is a function that checks whether the value in the datum is **nan** or not and returns either **True** or **False**. We use this condition to replace the values in the dataframe.

> **Tip**
> There are plenty of ways to manipulate and replace the values in a dataframe and H2O provides plenty of functionality to make implementation easy. Feel free to explore and experiment more with manipulating the values in the dataframe. You can find more details here: `https://docs.h2o.ai/h2o/latest-stable/h2o-py/docs/frame.html`.

Now that we have learned various methods to replace values in the dataframe, let's look into a more advanced approach to doing so that data scientists and engineers often take.

Imputation

Previously, we have seen how we can replace nan values in the dataset using `fillna()`, which sequentially replaces the nan data in the dataframe. The `fillna()` function fills data in a sequential manner; however, data need not always be sequential in nature. For example, consider a dataset of people buying gaming laptops. The dataset will mostly contain data about people in the age demographic of 13-28, with a few outliers. In such a scenario, if there are any nan values in the **age** column of the dataframe, then we cannot use the `fillna()` function to fill the nan values, as any nan value after any outlier value will introduce a bias in the dataframe. We need to replace the nan value with a value that is common among the standard distribution of the age group for that product, something that is between 13 and 28, rather than say 59, which is less likely.

Imputation is the process of replacing certain values in the dataframe with an appropriate substitute that does not introduce any bias or outliers that may impact model training. The method or formulas used to calculate the substitute value are termed the **imputation strategy**. Imputation is one of the most important methods of data processing, which handles missing and nan values and tries to replace them with a value that will potentially introduce the least bias into the model training process.

H2O has a function called `impute()` that specifically provides this functionality. It has the following parameters:

- `column`: This parameter accepts the column number that sets the columns to `impute()`. The value 1 imputes the entire dataframe.
- `method`: This parameter sets which method of imputation to use. The methods can be either `mean`, `median`, or `mode`.
- `combine_method`: This parameter dictates how to combine the quantiles for even samples when the imputation method chosen is `median`. The combination methods are either `interpolate`, `average`, `low`, or `high`.
- `group_by_frame`: This parameter imputes the values of the selected precomputed grouped frame.
- `by`: This parameter groups the imputation results by the selected columns.
- `values`: This parameter accepts a list of values that are imputed per column. Having the None value in the list skips the column.

Let's see an example in Python of how we can use this function to fill missing values.

For this, we shall use the **high school student sprint** dataset. The high school student sprint dataset is a dataset that consists of recordings of the age of high school students, their weight, maximum recorded speed, and their performance in a 100-meter sprint. The dataset is used to predict how the age, weight, and sprint speed affect the performance of students in a 100-meter sprint race.

The dataset looks as follows:

	age	weight	max_speed	100_meter_time
0	13	46	16.755530	NaN
1	15	33	15.089844	21.732242
2	13	32	14.584233	22.348161
3	16	45	15.669721	NaN
4	13	39	19.711957	20.169496
...	...	...	...	...
95	14	46	13.231185	NaN
96	16	45	17.572064	23.064310
97	17	38	14.639427	23.142284
98	17	32	15.984808	23.537361
99	14	48	16.072193	20.370838

100 rows × 4 columns

Figure 3.16 – A high school student sprint dataset

The features of the dataset are as follows:

- **age**: Age of the student
- **weight**: Weight of the student in kilograms
- **max_speed**: The maximum sprint speed of the student in kilometers per hour
- **100_meter_time**: The time taken by the student to finish a 100-meter sprint in seconds

As you can see, there are plenty of missing values in the **100_meter_time** column.

We cannot simply use the `fillna()` function, as that will introduce bias into the data if the missing values happen to be right after the fastest or slowest time. We can't simply replace the values with a constant number either.

What would actually make sense is to replace these missing values with whatever is normal for an average teenager doing a 100-meter dash. We already have values for the majority of students, so we can use their results to calculate a general average 100-meter sprint time and use that as a baseline to replace all the missing values without introducing any bias.

This is exactly what imputation is used for. Let's use the imputation function to fill in these missing values:

1. Import the h2o module and start the h2o server:

    ```
    import h2o
    h2o.init()
    ```

2. We then import the high school student sprint dataset by using h2o.import_file():

    ```
    dataframe = h2o.import_file(" Dataset/high_school_
    student_sprint.csv" )
    ```

3. Using the impute() function, let's impute the missing values in the 100_meter_time column by mean and display the data:

    ```
    dataframe.impute(" 100_meter_time" , method = " mean" )
    dataframe.describe
    ```

    You will see the output of the imputed dataframe as follows:

age	weight	max_speed	100_meter_time
13	41	17.9565	23.5558
16	45	13.6646	23.0656
15	44	18.1247	24.7465
15	39	14.621	23.7011
13	45	15.2694	25.8552
16	43	16.9983	24.6807
15	37	13.3174	24.7037
13	40	13.3024	21.8998
14	47	17.4593	26.3073
15	47	18.4853	22.5947

Figure 3.17 – 100_meter_time column imputed by its mean

4. H2O calculated the mean value of all the values in the **100_meter_time** column as **23.5558** and replaced the missing values with it.

Similarly, instead of mean, you can use median values as well. However, note that if a column has categorical values, then the method must be mode. The decision is up to you to make, depending on the dataset that is most useful when replacing the missing values:

```
dataframe.impute(" 100_meter_time" , method = " median" )
dataframe.impute(" 100_meter_time" , method = " mode" )
```

5. Let's increase the complexity a bit. What if the average 100-meter sprint time is not truly comparable between all students? What if the performances are more comparable age-wise? For example, students of age 16 are faster than the ones who are 13 since they are more physically developed. In that case, it won't make sense considering a 13-year-old's sprint time when imputing the missing value of a 16-year-old. This is where we can use the group parameter of the impute() function:

```
dataframe = h2o.import_file(" Dataset/high_school_
student_sprint.csv" )
dataframe.impute(" 100_meter_time" , method = " mean" ,
by=[" age" ])
dataframe.describe
```

You will see the output as follows:

age	weight	max_speed	100_meter_time
13	41	17.9565	22.9804
16	45	13.6646	23.0656
15	44	18.1247	24.7465
15	39	14.621	23.7011
13	45	15.2694	25.8552
16	43	16.9983	24.6807
15	37	13.3174	24.7037
13	40	13.3024	21.8998
14	47	17.4593	26.3073
15	47	18.4853	22.5947

Figure 3.18 – 100_meter_sprint imputed by its mean and grouped by age

You will notice that now H2O has calculated the mean values by age and replaced the respective missing values for that age in the **100_meter_time** column. Observe the first row in the dataset. The row was of students aged 13 and had missing values in its **100_meter_time** column. It was replaced with the mean value of all the **100_meter_time** values for other 13-year-olds. Similar steps were followed for other age groups. This is how you can use the group by parameter in the impute() function to flexibly impute the correct values.

The impute() function is extremely powerful to impute the correct values in a dataframe. The additional parameters for grouping via columns as well as frames make it very flexible for use in handling all sorts of missing values.

Feel free to use and explore all these functions on different datasets. At the end of the day, all these functions are just tools used by data scientists and engineers to improve the quality of the data; the real skill is understanding when and how to use these tools to get the most out of your data, and that requires experimentation and practice.

Now that we have learned about the different ways in which we can handle missing data, let's move on to the next part of data processing, which is how to manipulate the feature columns of the dataframe.

Manipulating feature columns of the dataframe

The majority of the time, your data processing activities will mostly involve manipulating the columns of the dataframes. Most importantly, the type of values in the column and the ordering of the values in the column will play a major role in model training.

H2O provides some functionalities that help you do so. The following are some of the functionalities that help you handle missing values in your dataframe:

- Sorting of columns
- Changing the type of the column

Let's first understand how we can sort a column using H2O.

Sorting columns

Ideally, you want the data in a dataframe to be shuffled before passing it off to model training. However, there may be certain scenarios where you might want to re-order the dataframe based on the values in a column.

H2O has a functionality called `sort()` to sort dataframes based on the values in a column. It has the following parameters:

- by: The column to sort by. You can pass multiple column names as a list as well.

- `ascending`: A `boolean` array that denotes the direction in which H2O should sort the columns. If `True`, H2O will sort the column in ascending order. If `False`, then H2O will sort it in descending order. If neither of the flags is passed, then H2O defaults to sorting in ascending order.

The way H2O will sort the dataframe depends on whether one column name is passed to the `sort()` function or multiple column names. If only a single column name is passed, then H2O will return a frame that is sorted by that column.

However, if multiple columns are passed, then H2O will return a dataframe that is sorted as follows:

- H2O will first sort the dataframe on the first column that is passed in the parameter.

- H2O will then sort the dataframe on the next column passed in the parameter, but only those rows will be sorted that have the same values as in the first sorted column. If there are no duplicate values in the previous columns, then no sorting will be done on subsequent columns.

Let's see an example in Python of how we can use this function to sort columns:

1. Import the h2o library and initialize it:

```
import h2o
h2o.init()
```

2. Create a dataframe by executing the following code and observe the dataset:

```
dataframe = h2o.H2OFrame.from_python({'C1':
[3,3,3,0,12,13,1,8,8,14,15,2,3,8,8],'C2':
[1,5,3,6,8,6,8,7,6,5,1,2,3,6,6]
,'C3':[15,14,13,12,11,10,9,8,7,6,5,4,3,2,1]})
dataframe.describe
```

The contents of the dataset should be as follows:

	C1	C2	C3
0	6	12	
1	8	9	
2	2	4	
3	1	15	
3	5	14	
3	3	13	
3	3	3	
8	7	8	
8	6	7	
8	6	2	

Figure 3.19 – dataframe_1 data contents

3. So, at the moment, the values in columns **C1**, **C2**, and **C3** are all random in nature. Let's use the sort () function to sort the dataframe by column **C1**. You can do so by either passing 0 into the by parameter, indicating the first column of the dataframe, or by passing ['C1'], which is a list containing column names to sequentially sort the dataset:

```
sorted_dataframe_1 = dataframe.sort(0)
sorted_dataframe_1.describe
```

You should get an output of the code as follows:

	C1	C2	C3
0	6	12	
1	8	9	
2	2	4	
3	1	15	
3	5	14	
3	3	13	
3	3	3	
8	7	8	
8	6	7	
8	6	2	

Figure 3.20 – dataframe_1 sorted by the C1 column

You will see that the dataframe is now sorted in ascending order by the **C1** column.

4. Let's see what we shall get if we pass multiple columns in the by parameter to sort on multiple columns. Run the following code line:

```
sorted_dataframe_2 = dataframe.sort(['C1','C2']) sorted_
dataframe_2.describe
```

You should get an output as follows:

C1	C2	C3
0	6	12
1	8	9
2	2	4
3	1	15
3	3	13
3	3	3
3	5	14
8	6	7
8	6	2
8	6	1

Figure 3.21 – dataframe_1 sorted by columns C1 and C2

As you can see, H2O first sorted the columns by the **C1** column. Then, it sorted the rows by the **C2** column for those rows that had the same value in the **C1** column. H2O will sequentially sort the dataframe column-wise for all the columns you pass in the sort function.

5. You can also reverse the sorting order by passing False in the ascending parameter. Let's test this out by running the following code line:

```
sorted_dataframe_3 = dataframe.sort(by=['C1','C2'],
ascending=[True,False])
sorted_dataframe_3.describe
```

You should see an output as follows:

C1	C2	C3
0	6	12
1	8	9
2	2	4
3	5	14
3	3	13
3	3	3
3	1	15
8	7	8
8	6	7
8	6	2

Figure 3.22 – dataframe_1 sorted by the C1 column in ascending order and the C2 column in descending order

In this case, H2O first sorted the columns by the **C1** column. Then, it sorted the rows by the **C2** column for those rows that had the same value in the **C1** column. However, this time it sorted the values in descending order.

Now that you've learned how to sort the dataframe by a single column as well as by multiple columns, let's move on to another column manipulation function that changes the type of the column.

Changing column types

As we saw in *Chapter 2, Working with H2O Flow (H2O's Web UI)*, we changed the type of the Heart Disease column to enum from numerical. The reason we did this is that the type of column plays a major role in model training. During model training, the type of column decides whether the ML problem is a classification problem or a regression problem. Despite the fact that the data in both cases is numerical in nature, how a ML algorithm will treat the column depends entirely on its type. Thus, it becomes very important to correct the types of columns that might not be correctly set during the initial stages of data collection.

H2O has several functions that not only help you change the type of the columns but also run initial checks on the column types.

Some of the functions are as follows:

- `.isnumeric()`: Checks whether the column in the dataframe is of the numeric type. Returns `True` or `False` accordingly

- `.asnumeric()`: Creates a new frame with all the values converted to numeric for the specified column

- `.isfactor()`: Checks whether the column in the dataframe is of categorical type. Returns `True` or `False` accordingly

- `.asfactor()`: Creates a new frame with all the values converted to the categorical type for the specified column

- `.isstring()`: Checks whether the column in the dataframe is of the string type. Returns `True` or `False` accordingly

- `.ascharacter()`: Creates a new frame with all the values converted to the string type for the specified column

Let's see an example in Python of how we can use these functions to change the column types:

1. Import the h2o library and initialize H2O:

   ```
   import h2o
   h2o.init()
   ```

2. Create a dataframe by executing the following code line and observe the dataset:

   ```
   dataframe = h2o.H2OFrame.from_python({'C1':
   [3,3,3,0,12,13,1,8,8,14,15,2,3,8,8],'C2':
   [1,5,3,6,8,6,8,7,6,5,1,2,3,6,6]
   ,'C3':[15,14,13,12,11,10,9,8,7,6,5,4,3,2,1]})
   dataframe.describe
   ```

The contents of the dataset should be as follows:

C1	C2	C3
3	1	15
3	5	14
3	3	13
0	6	12
12	8	11
13	6	10
1	8	9
8	7	8
8	6	7
14	5	6

Figure 3.23 – Dataframe data contents

3. Let's confirm whether the **C1** column is a numerical column by using the `isnumeric()` function as follows:

    ```
    dataframe['C1'].isnumeric()
    ```

    You should get an output of `True`.

4. Let's see what we get if we check whether the **C1** column is a categorical column using the `asfactor()` function as follows:

    ```
    dataframe['C1'].isfactor()
    ```

    You should get an output of `False`.

5. Now let's convert the **C1** column to a categorical column using the `asfactor()` function and then check whether `isfactor()` returns `True`:

    ```
    dataframe['C1'] = dataframe['C1'].asfactor()
    dataframe['C1'].isfactor()
    ```

    You should now get an output of `True`.

6. You can convert the **C1** column back into a numerical column by using the `asnumeric()` function:

    ```
    dataframe['C1'] = dataframe['C1'].asnumeric()
    dataframe['C1'].isnumeric()
    ```

    You should now get an output of `True`.

Now that you have learned how to sort the columns of a dataframe and change column types, let's move on to another important topic in data processing, which is tokenization and encoding.

Tokenization of textual data

Not all **Machine Learning Algorithms** (**MLAs**) are focused on mathematical problem-solving. **Natural Language Processing** (**NLP**) is a branch of ML that specializes in analyzing meaning out of textual data, though it will try to derive meaning and understand the contents of a document or any text for that matter. Training an NLP model can be very tricky, as every language has its own grammatical rules and the interpretation of certain words depends heavily on context. Nevertheless, an NLP algorithm often tries its best to train a model that can predict the meaning and sentiments of a textual document.

The way to train an NLP algorithm is to first break down the chunk of textual data into smaller units called **tokens**. Tokens can be words, characters, or even letters. It depends on what the requirements of the MLA are and how it uses these tokens to train a model.

H2O has a function called `tokenize()` that helps break down string data in a dataframe into tokens and creates a separate column containing all the tokens for further processing.

It has the following parameter: `split`: We pass a regular expression in this parameter that will be used by the function to split the text data into tokens.

Let's see an example of how we can use this function to tokenize string data in a dataframe:

1. Import the h2o library and initialize it:

    ```
    import h2o
    h2o.init()
    ```

2. Create a dataframe by executing the following code line and observe the dataset:

    ```
    dataframe1 = h2o.H2OFrame.from_python({'C1':['Today we
    learn AI', 'Tomorrow AI learns us', 'Today and Tomorrow
    are same', 'Us and AI are same']})
    dataframe1 = dataframe1.ascharacter()
    dataframe1.describe
    ```

The dataset should look as follows:

C1
Today we learn AI
Tomorrow AI learns us
Today and Tomorrow are same
Us and AI are same

Figure 3.24 – Dataframe data contents

This type of textual data is usually collected in systems that generate a lot of log text or conversational data. To solve such NLP tasks, we need to break down the sentences into individual tokens so that we can eventually build the context and meaning of these texts that will help the ML algorithm to make semantic predictions. However, before diving into the complexities of NLP, data scientists and engineers will process this data by tokenizing it first.

3. Let's tokenize our dataframe using this function to split the text with blank spaces and observe the tokenized column:

```
tokenized_dataframe = dataframe1.tokenize("   ")
tokenized_dataframe
```

You should see the dataframe as follows:

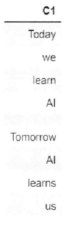

C1
Today
we
learn
AI
Tomorrow
AI
learns
us

Figure 3.25 – Tokenized dataframe data contents

You will notice that the `tokenize()` function splits the text data into tokens and appends the tokens as rows into a single column. You will also notice that all tokenized sentences are separated by empty rows. You can cross-check this by comparing the number of words in all the sentences in the dataframe, plus the empty spaces between the sentences against the number of rows in the tokenized dataset, using `nrows`.

These are some of the most used data processing methods that are used to process your data before you feed it to your ML pipeline for training. There are still plenty of methods and techniques that you can use to further clean and polish your dataframes. So much so that you could dedicate an entire book to discussing them. Data processing happens to be the most difficult part of the entire ML life cycle. The quality of the data used for training depends on the context of the problem statement. It also depends on the creativity and ingenuity of the data scientists and engineers in processing that data. The end goal of data processing is to extract as much information as we can from the dataset and remove noise and bias from the data to allow for a more efficient analysis of data during training.

Encoding data using target encoding

As we know, machines are only capable of understanding numbers. However, plenty of real-world ML problems revolve around objects and information that are not necessarily numerical in nature. Things such as states, names, and classes, in general, are represented as categories rather than numbers. This kind of data is called **categorical data**. Categorical data will often play a big part in analysis and prediction. Hence, there is a need to convert these categorical values to a numerical format so that machines can understand them. The conversion should also be in such a way that we do not lose the inherent meaning of those categories, nor do we introduce new information into the data, such as the incremental nature of numbers, for example.

This is where encoding is used. **Encoding** is a process where categorical values are transformed, in other words, *encoded*, into numerical values. There are plenty of encoding methods that can perform this transformation. One of the most commonly used ones is **target encoding**.

Target encoding is an encoding process that transforms categorical values into numerical values by calculating the average probability of the target variable occurring for a given category. H2O also has methods that help users implement target encoding on their data.

To better understand this method, consider the following sample `Mythical creatures` dataset:

Animals	Mythical
Dragon	0
Unicorn	1
Horse	1
Lizard	1
Goblin	0
Dragon	0
Horse	0
Horse	1
Unicorn	0
Dragon	1
Goblin	0
Lizard	1
Lizard	1
Unicorn	0
Dragon	1

Figure 3.26 – Our mythical creatures dataset

This dataset has the following content:

- **Animals**: This column contains categorical values of the names of animals.
- **Mythical**: This column contains the **0** binary value and the **1** binary value. **1** indicates that the creature is mythical, while **0** indicates that the creature is not mythical.

Now, let's encode the **Animals** `categorical` column using target encoding. Target encoding will perform the following steps:

1. Group the categorical values and record the number of times the target value, **Mythical**, was 1 and when it was 0 for a given category as follows:

Animals_target_encoded	Target 0 count	Target 1 count
Dragon	2	2
Unicorn	2	1
Horse	1	2
Lizard	0	3
Goblin	2	0

Figure 3.27 – The mythical creatures dataset with a target count

2. Calculate the probability that the **1** target value will occur, as compared to the **0** target value within each specific group. This would look as follows:

Animals_target_encoded	Target 0 count	Target 1 count	Probability of Target 1 Occurring
Dragon	2	2	0.50
Unicorn	2	1	0.33
Horse	1	2	0.66
Lizard	0	3	1
Goblin	2	0	0

Figure 3.28 – The mythical creatures dataset with a Probability of Target 1 Occurring column

3. Drop the **Animals** column and use the **Probability of Target 1 Occurring** column as the encoded representation of the **Animals** column. The new encoded dataset will look as follows:

Animals	Mythical
0.50	0
0.33	1
0.66	1
1	0
0	1
0.50	0
0.66	0
0.66	1
0.33	0
0.50	1
0	0
1	1
1	1
0.33	0
0.50	1

Figure 3.29 – A target-encoded mythical creatures dataset

In the encoded dataset, the **Animals** feature is encoded using target encoding and we have a dataset that is entirely numerical in nature. This dataset will be easy for an ML algorithm to interpret and learn from, providing high-quality models.

Let us now see how we can perform target encoding using H2O. The dataset we will use for this example is the Automobile price prediction dataset. You can find the details of this dataset at https://archive.ics.uci.edu/ml/datasets/Automobile (*Dua, D. and Graff, C. (2019). UCI Machine Learning Repository* [http://archive.ics.uci.edu/ml]. *Irvine, CA: University of California, School of Information and Computer Science*).

The dataset is fairly straightforward. It contains various details about cars, such as the **make of the car**, **engine size**, **fuel system**, **compression ratio**, and **price**. The aim of the ML algorithm is to predict the price of a car based on these features.

For our experiment, we shall encode the categorical columns **make**, **fuel type**, and **body style** using target encoding where the **price** column is the target.

Let's perform target encoding by following this example:

1. Import h2o and H2O's **target encoder** library, H2OTargetEncoderEstimator, and initialize your H2O server. Execute the following code:

```
import h2o
from h2o.estimators import H2OTargetEncoderEstimator
h2o.init()
```

2. Import the Automobile price prediction dataset and print the contents of the dataset. Execute the following code:

```
automobile_dataframe = h2o.import_file(" Dataset\
Automobile_data.csv" )
automobile_dataframe
```

Let's observe the contents of the dataframe; it should look as follows:

Parse progress: |██| (done) 100%

symboling	normalized-losses	make	fuel-type	aspiration	num-of-doors	body-style	drive-wheels	engine-location	wheel-base	length	width	height	curb-weight	engine-type	num-of-cylinders	engine-size	fuel-system
3	nan	alfa-romero	gas	std	two	convertible	rwd	front	88.6	168.8	64.1	48.8	2548	dohc	four	130	mpi
3	nan	alfa-romero	gas	std	two	convertible	rwd	front	88.6	168.8	64.1	48.8	2548	dohc	four	130	mpi
1	nan	alfa-romero	gas	std	two	hatchback	rwd	front	94.5	171.2	65.5	52.4	2823	ohcv	six	152	mpi
2	164	audi	gas	std	four	sedan	fwd	front	99.8	176.6	66.2	54.3	2337	ohc	four	109	mpi
2	164	audi	gas	std	four	sedan	4wd	front	99.4	176.6	66.4	54.3	2824	ohc	five	136	mpi
2	nan	audi	gas	std	two	sedan	fwd	front	99.8	177.3	66.3	53.1	2507	ohc	five	136	mpi
1	158	audi	gas	std	four	sedan	fwd	front	105.8	192.7	71.4	55.7	2844	ohc	five	136	mpi
1	nan	audi	gas	std	four	wagon	fwd	front	105.8	192.7	71.4	55.7	2954	ohc	five	136	mpi
1	158	audi	gas	turbo	four	sedan	fwd	front	105.8	192.7	71.4	55.9	3086	ohc	five	131	mpi
0	nan	audi	gas	turbo	two	hatchback	4wd	front	99.5	178.2	67.9	52	3053	ohc	five	131	mpi

Figure 3.30 – An automobile price prediction dataframe

As you can see in the preceding figure, the dataframe consists of a large number of columns containing the details of cars. For the sake of understanding target encoding, let's filter out the columns that we want to experiment with while dropping the rest. Since we plan on encoding the make column, the fuel-type column, and the body-style column, let's use only those columns along with the price response column. Execute the following code:

```
automobile_dataframe = automobile_dataframe[:, [" make" ,
" fuel-type" , " body-style" , " price" ]]
automobile_dataframe
```

The filtered dataframe will look as follows:

make	fuel-type	body-style	price
alfa-romero	gas	convertible	13495
alfa-romero	gas	convertible	16500
alfa-romero	gas	hatchback	16500
audi	gas	sedan	13950
audi	gas	sedan	17450
audi	gas	sedan	15250
audi	gas	sedan	17710
audi	gas	wagon	18920
audi	gas	sedan	23875
audi	gas	hatchback	nan

Figure 3.31 – The automobile price prediction dataframe with filtered columns

3. Let's now split this dataframe into training and testing dataframes. Execute the following code:

```
automobile_dataframe_for_training, automobile_dataframe_
for_test = automobile_dataframe.split_frame(ratios =
[.8], seed = 123)
```

4. Let's now train our target encoder model using H2OTargetEncoderEstimator. Execute the following code:

```
automobile_te = H2OTargetEncoderEstimator()
automobile_te.train(x= [" make" , " fuel-type" , "
body-style" ], y=" price" , training_frame=automobile_
dataframe_for_training)
```

Once the target encoder has finished its training, you will see the following output:

```
targetencoder Model Build progress: |████████████████████████████████████████| (done) 100%
Model Details
=============
H2OTargetEncoderEstimator :  TargetEncoder
Model Key:  TargetEncoder_model_python_1653850929913_2


Target Encoder model summary: Summary for target encoder model
```

	original_names	encoded_column_names
0	make	make_te
1	fuel-type	fuel-type_te
2	body-style	body-style_te

Figure 3.32 – The result of target encoder training

From the preceding screenshot, you can see that the H2O target encoder will generate the target-encoded values for the make column, the fuel-type column, and the body-style column and store them in different columns named make_te, fuel-type_te, and body-style_te, respectively. These new columns will contain the encoded values.

5. Let's now use this trained target encoder to encode the training dataset and print the encoded dataframe:

```
te_automobile_dataframe_for_training = automobile_
te.transform(frame=automobile_dataframe_for_training, as_
training=True)
te_automobile_dataframe_for_training
```

The encoded training frame should look as follows:

make_te	fuel-type_te	body-style_te	make	fuel-type	body-style	price
15498.3	13074	21890.5	alfa-romero	gas	convertible	13495
15498.3	13074.1	21890.5	alfa-romero	gas	convertible	16500
15498.3	13074.1	9722.24	alfa-romero	gas	hatchback	16500
16656	13074.1	14734.1	audi	gas	sedan	13950
16656	13074.1	14734.1	audi	gas	sedan	17450
16656	13074.1	14734.1	audi	gas	sedan	15250
16656	13074.1	14734.1	audi	gas	sedan	17710
16656	13074.1	11809.9	audi	gas	wagon	18920
26340.7	13074.1	14734.1	bmw	gas	sedan	16430
26340.7	13074.1	14734.1	bmw	gas	sedan	16925

Figure 3.33 – An encoded automobile price prediction training dataframe

As you can see from the figure, our training frame now has three additional columns, make_te, fuel-type_te, and body-style_te, with numerical values. These are the target-encoded columns for the make column, the fuel-type column, and the body-style column.

6. Similarly, let's now use the trained target encoder to encode the test dataframe and print the encoded dataframe. Execute the following code:

```
te_automobile_dataframe_for_test = automobile_
te.transform(frame=automobile_dataframe_for_test,
noise=0)
te_automobile_dataframe_for_test
```

The encoded test frame should look as follows:

make_te	fuel-type_te	body-style_te	make	fuel-type	body-style	price
16656	13074.1	14734.1	audi	gas	sedan	23875
16656	13074.1	9722.23	audi	gas	hatchback	nan
26340.7	13074.1	14734.1	bmw	gas	sedan	24565
7790.12	13074.1	14734.1	dodge	gas	sedan	8558
8151	13074.1	9722.23	honda	gas	hatchback	7895
8151	13074.1	14734.1	honda	gas	sedan	8845
11048	13074.1	14734.1	isuzu	gas	sedan	6785
34125	13074.1	14734.1	jaguar	gas	sedan	35550
11260.3	13074.1	9722.23	mazda	gas	hatchback	6795
11260.3	13074.1	9722.23	mazda	gas	hatchback	8845

Figure 3.34 – An encoded automobile price prediction test dataframe

As you can see from the figure, our test frame also has three additional columns, which are the encoded columns. You can now use these dataframes to train your ML models.

Depending on your next actions, you can use the encoded dataframes however you see fit. If you want to use the dataframe to train ML models, then you can drop the `categorical` columns from the dataframe and use the respective encoded columns as training features to train your models. If you wish to perform any further analytics on the dataset, then you can keep both types of columns and perform any comparative study.

> **Tip**
>
> H2O's target encoder has several parameters that you can set to tweak the encoding process. Selecting the correct settings for target encoding your dataset can get very complex, depending on the type of data with which you are working. So, feel free to experiment with this function, as the better you understand this feature and target encoding in general, the better you can encode your dataframe and further improve your model training. You can find more details about H2O's target encoder here: `https://docs.h2o.ai/h2o/latest-stable/h2o-docs/data-science/target-encoding.html`.

Congratulations! You have just understood how you can encode categorical values using H2O's target encoder.

Summary

In this chapter, we first explored the various techniques and some of the common functions we use to preprocess our dataframe before it is sent to model training. We looked into how we can reframe our raw dataframe into a suitable consistent format that meets the requirement for model training. We learned how to manipulate the columns of dataframes by combining them with different columns of different dataframes. We learned how to combine rows from partitioned dataframes, as well as how to directly merge dataframes into a single dataframe.

Once we knew how to reframe our dataframes, we learned how to handle the missing values that are often present in freshly collected data. We learned how to fill NA values, replace certain incorrect values, as well as how to use different imputation strategies to avoid adding noise and bias when filling missing values.

We then investigated how we can manipulate the feature columns by sorting the dataframes by column, as well as changing the types of columns. We also learned how to tokenize strings to handle textual data, as well as how to encode categorical values using H2O's target encoder.

In the next chapter, we will open the black box of AutoML, explore its training, and what happens internally during the AutoML process. This will help us to better understand how H2O does its magic and efficiently automates the model training process.

Understanding H2O AutoML Architecture and Training

Model training is one of the core components of a **Machine Learning** (**ML**) pipeline. It is the step in the pipeline where the system reads and understands the patterns in the dataset. This learning outputs a mathematical representation of the relationship between the different features in the dataset and the target value. The way in which the system reads and analyzes data depends on the ML algorithm being used and its intricacies. This is where the primary complexity of ML lies. Every ML algorithm has its own way of interpreting the data and deriving information from it. Every ML algorithm aims to optimize certain metrics while trading off certain biases and variances. Automation done by H2O AutoML further complicates this concept. Trying to understand how that would work can be overwhelming for many engineers.

Don't be discouraged by this complexity. All sophisticated systems can be broken down into simple components. Understanding these components and their interaction with each other is what helps us understand the system as a whole. Similarly, in this chapter, we will open up the black box, that is, H2O's AutoML service, and try to understand what kind of magic makes the automation of ML possible. We shall first understand the architecture of H2O. We shall break it down into simple components and then understand what interaction takes place between the various components of H2O. Later, we will come to understand how H2O AutoML trains so many models and is able to optimize their hyperparameters to get the best possible model.

In this chapter, we are going to cover the following topics:

- Observing the high-level architecture of H2O
- Knowing the flow of interaction between the client and the H2O service
- Understanding how H2O AutoML performs hyperparameter optimization and training

So, let's begin by first understanding the architecture of H2O.

Observing the high-level architecture of H2O

To deep dive into H2O technology, we first need to understand its high-level architecture. It will not only help us understand what the different software components that make up the H2O AI stack are, but it will also help us understand how the components interact with each other and their dependencies.

With this in mind, let's have a look at the H2O AI high-level architecture, as shown in the following diagram:

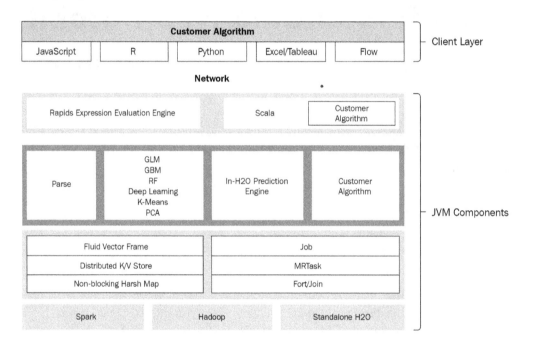

Figure 4.1 – H2O AI high-level architecture

The H2O AI architecture is conceptually divided into two parts, each serving a different purpose in the software stack. The parts are as follows:

- **Client layer** – This layer points to the client code that communicates with the H2O server.
- **Java Virtual Machine (JVM) components** – This layer indicates the H2O server and all of its JVM components that are responsible for the different functionalities of H2O AI, including AutoML.

The client and the JVM component layers are separated by the **network layer**. The network layer is nothing but the general internet, which requests are sent over.

Let's dive deep into every layer to better understand their functionalities, starting with the first layer, the client layer.

Observing the client layer

The client layer comprises all the client code that you install in your system. You use this software program to send requests to the H2O server to perform your ML activities. The following diagram shows you the client layer from the H2O high-level architecture:

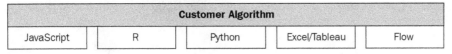

Figure 4.2 – The client layer of H2O high-level architecture

Every supported language will have its own H2O client code that is installed and used in the respective language's script. All client code internally communicates with the H2O server via a REST API over a socket connection.

The following H2O clients exist for the respective languages:

- **JavaScript**: H2O's embedded web UI is written in JavaScript. When you start the H2O server, it starts a JavaScript web client that is hosted on `http://localhost:54321`. You can log into this client with your web browser and communicate with the H2O server to perform your ML activities. The JavaScript client communicates with the H2O server via a REST API.

- **R**: Referring to *Chapter 1, Understanding H2O AutoML Basics*, we import the H2O library by executing `library(h2o)` and then using the imported H2O variable to import the dataset and train models. This is the R client that is interacting with the initialized H2O server and it does so using a REST API.

- **Python**: Similarly, in *Chapter 1, Understanding H2O AutoML Basics*, we import the H2O library in Python by executing `import h2o` and then using the imported H2O variable to command the H2O server. This is the Python client that is interacting with the H2O server using a REST API.

- **Excel**: Microsoft Excel is spreadsheet software developed by Microsoft for Windows, macOS, Android, and iOS. H2O has support for Microsoft Excel as well since it is the most widely used spreadsheet software that handles large amounts of two-dimensional data. This data is well suited for analytics and ML. There is an H2O client for Microsoft Excel as well that enables Excel users to use H2O for ML activities through the Excel client.

- **Tableau**: Tableau is interactive data visualization software that helps data analysts and scientists visualize data in the form of graphs and charts that are interactive in nature. H2O has support for Tableau and, as such, has a dedicated client for Tableau that adds ML capabilities to the data ingested by Tableau.

- **Flow**: As seen in *Chapter 2, Working with H2O Flow (H2O's Web UI)*, H2O Flow is H2O's web user interface that has all the functional capabilities of setting up the entire ML lifecycle in a notebook-style interface. This interface internally runs on JavaScript and similarly communicates with the H2O server via a standard REST API.

The following diagram shows you the interactions of various H2O clients with the same H2O server:

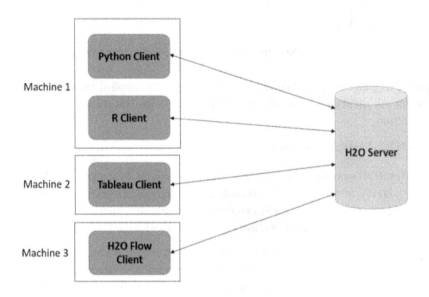

Figure 4.3 – Different clients communicating with the same H2O server

As you can see in the diagram, all the different clients can communicate with the same instance of the H2O server. This enables a single H2O server to service different software products written in different languages.

This covers the contents of the client layer; let's move down to the next layer in the H2O's high-level architecture, that is, the JVM component layer.

Observing the JVM component layer

The JVM is a runtime engine that runs Java programs in your system. The H2O cloud server runs on multiple **JVM processes**, also called **JVM nodes**. Each JVM node runs specific components of the H2O software stack.

The following diagram shows you the various JVM components that make up the H2O server:

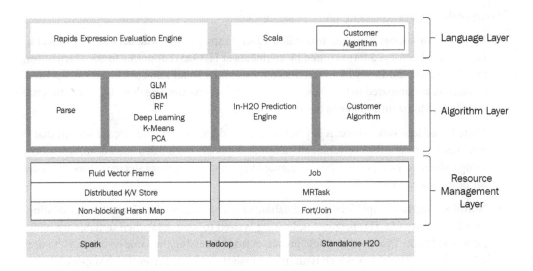

Figure 4.4 – H2O JVM component layer

As seen in the preceding diagram, the JVM nodes are further split into three different layers, which are as follows:

- **The language layer**: This layer contains processes that are responsible for evaluating the different client language expressions that are sent to the H2O cloud server. The R evaluation layer is a slave to its respective REST client. This layer also contains the Shalala Scala layer. The `Shalala Scala` library is a code library that accesses dedicated domain-specific language that users can use to write their own programs and algorithms that H2O can use.

- **The algorithm layer**: This layer contains all the inbuilt ML algorithms that H2O provides. The JVM processes in this layer are responsible for performing all the ML activities, such as importing datasets, parsing, calculating the mathematics for the respective ML algorithms, and overall training of models. This layer also has the prediction engine whose processes perform prediction and scoring functions using the trained models. Any custom algorithms imported into H2O are also housed in this layer and the JVM processes handle the execution just like the other algorithms.

- **The resource management layer**: This layer contains all the JVM processes responsible for the efficient management of system resources such as memory and CPU when performing ML activities.

Some of the JVM processes in this layer are as follows:

- **Fluid Vector frame**: A frame, also called a DataFrame, is the basic data storage object in H2O. Fluid Vector is a term coined by engineers at H2O.ai that points to the efficient (or, in other words, fluid) way by which the columns in the DataFrame can be added, updated, or deleted, as compared to the DataFrames in the data engineering domain, where they are usually said to be immutable in nature.

- **Distributed key-value store**: A key-value store or database is a data storage system that is designed to retrieve data or values efficiently and quickly from a distributed storage system using indexed keys. H2O uses this distributed key-value in-memory storage across its cluster for quick storage and lookups.

- **NonBlockingHashMap**: Usually in a database to provide **Atomicity, Consistency, Isolation, and Durability** (**ACID**) properties, locking is used to lock data when updates are being performed on it. This stops multiple processes from accessing the same resource. H2O uses a NonBlockingHashMap, which is an implementation of ConcurrentHashMap with better scaling capabilities.

- **Job**: In programming, a job is nothing but a large piece of work that is done by software that serves a single purpose. H2O uses a jobs manager that orchestrates various jobs that perform complex tasks such as mathematical computations with increased efficiency and less CPU resource consumption.

- **MRTask**: H2O uses its own in-memory MapReduce task to perform its ML activities. MapReduce is a programming model that is used to process large amounts of computation or data read and writes using parallel execution of tasks on a distributed cluster. MapReduce helps the system perform computational activities faster than sequential computing.

- **Fork/Join**: H2O uses a modified Java concurrency library called **jsr166y** to perform the concurrent execution of tasks. jsr166y is a very lightweight task execution framework that uses **Fork**, where the process breaks down a task into smaller subtasks, and **Join**, where the process joins the results of the subtasks together to get the final output of the task.

The entire JVM component layer lies on top of **Spark** and **Hadoop** data processing systems. The components in the JVM layer leverage these data processing cluster management engines to support cluster computing.

This sums up the entire high-level architecture of H2O's software technology. With this background in mind, let's move to the next section, where we shall understand the flow of interaction between the client and H2O and how the client-server interaction helps us perform ML activities.

Learning about the flow of interaction between the client and the H2O service

In *Chapter 1*, *Understanding H2O AutoML Basics*, and *Chapter 2*, *Working with H2O Flow (H2O's Web UI)*, we saw how we can send a command to H2O to import a dataset or train a model. Let's try to understand what happens behind the scenes when you send a request to the H2O server, beginning with data ingestion.

Learning about H2O client-server interactions during the ingestion of data

The process of a system ingesting data is the same as how we read a book in real life: we open the book and start reading one line at a time. Similarly, when you want your program to read a dataset stored in your system, you will first inform the program about the location of the dataset. The program will then open the file and start reading the bytes of the data line by line and store it in its RAM. However, the issue with the type of sequential data reading in ML is that datasets tend to be huge in ML. Such data is often termed big data and can span from gigabytes to terabytes of volume. Reading such huge volumes of data by a system, no matter how fast it may be, will need a significant amount of time. This is time that ML pipelines do not have, as the aim of an ML pipeline is to make predictions. These predictions won't have any value if the time to make decisions has already passed. For example, if you design an ML system that is installed in a car that automatically stops the car if it detects a possibility of collision, then the ML system would be useless if it spent all its time reading data and was too late to make collision predictions before they happened.

This is where **parallel computing** or **cluster computing** comes in. A **cluster** is nothing but multiple processes connected together over a network that performs like a single entity. The main aim of cluster computing is to parallelize long-running sequential tasks using these multiple processes to finish the task quickly. It is for this reason that cluster computing plays a very important role in ML pipelines. H2O also rightly uses clusters to ingest data.

Let's observe how a data ingestion interaction request flows from the H2O client to the H2O server and how H2O ingests data.

Refer to the following diagram to understand the flow of data ingestion interaction:

Figure 4.5 – H2O data ingestion request interaction flow

The following sequence of steps describes how a client request to the H2O cluster server to ingest data is serviced by H2O using the **Hadoop Distributed File System (HDFS)**:

1. **Making the request**: Once the H2O cluster server is up and running, the user using the H2O client makes a data ingestion function call pointing to the location of the dataset (see **Step 1** in *Figure 4.5*). The function call in Python would be as follows:

    ```
    h2o.import_file("Dataset/iris.data")
    ```

    The H2O client will extract the dataset location from the function call and internally create a REST API request (see **Step 2** in *Figure 4.5*). The client will then send the request over the network to the IP address where the H2O server is hosted.

2. **H2O server processing the request**: Once the H2O cluster server receives the HTTP request from the client, it will extract the dataset location path value from the request and initiate the distributed dataset ingestion process (see **Step 3** in *Figure 4.5*). The cluster nodes will then coordinate and parallelize the task of reading the dataset from the given path (see **Step 4** in *Figure 4.5*).

    Each node will read a section of the dataset and store it in its cluster memory.

3. **Ingestion of data**: The data read from the dataset location path will be stored in blocks in the distributed H2OFrame cluster memory (see **Step 1** in *Figure 4.6*). The block of data is stored in a distributed key-value store (see **Step 2** in *Figure 4.6*). Once the data is fully ingested, the

H2O server will create a pointer that points to the ingested dataset stored in the key-value store and return it to the requesting client (see **Step 3** in *Figure 4.6*).

Refer to the following diagram to understand the flow of interaction once data is ingested and H2O returns a response:

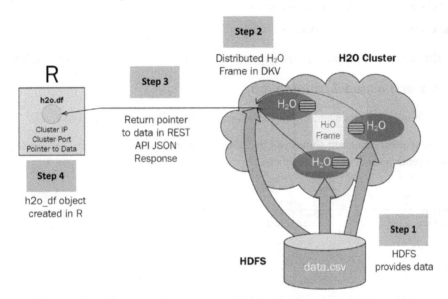

Figure 4.6 – H2O data ingestion response interaction flow

Once the client receives the response, it creates a DataFrame object that contains this pointer, which the user can then later use to run any further executions on the ingested dataset (see **Step 4** in *Figure 4.6*). In this way, with the use of pointers and the distributed key-value store, H2O can work on DataFrame manipulations and usage without needing to transfer the huge volume of data that it ingested between the server and client.

Now that we understand how H2O ingests data, let us now look into how it handles model training requests.

Knowing the sequence of interactions in H2O during model training

During model training, there are plenty of interactions that take place, right from the users making the model training request to the user getting the trained ML model. The various components of H2O perform the model training activity using a series of coordinated messages and scheduled jobs.

To better understand what happens internally when a model training request is sent to the H2O server, we need to dive deep into the sequence of interactions that occur during model training.

We shall understand the sequences of interactions by categorizing them as follows:

1. The client starts a model training job.

2. H2O runs the model training job.

3. The client polls for job completion status.

4. The client queries for the model information.

So, let's begin first by understanding what happens when the client starts a model training job.

The client starts a model training job

The model training job starts when the client first sends a model training request to H2O.

The following sequence diagram shows you the sequence of interactions that take place inside H2O when a client sends a model training request:

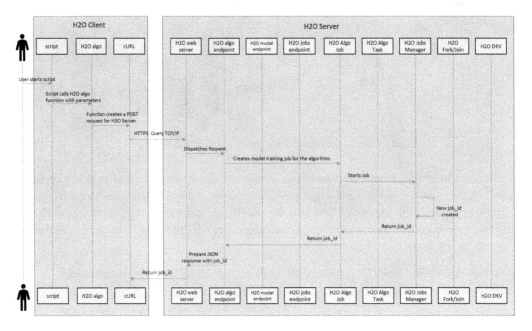

Figure 4.7 – Sequence of interactions in the model training request

The following set of sequences takes place during a model training request:

1. The user first runs a script that contains all the instructions and function calls to make a model training request to H2O.

2. The script contains a model training function call with its respective parameters. This also includes an H2O AutoML function call that performs in a similar manner.

3. The function call instructs the respective language-specific H2O client, which creates a **POST** request that contains all the parametric information needed to train the model correctly.

4. The H2O client will then perform a **curl** operation that sends the HTTP POST request to the H2O web server at the host IP address that it is hosted on.

5. From this point onward, the flow of information is performed inside the H2O server. The H2O server dispatches the request to the appropriate model training endpoint based on the model that was chosen to be trained by the user.

6. This model training endpoint extracts the parameter values from the request and schedules a job.

7. The job, once scheduled, starts training the model.

8. The **job manager** is responsible for handling all the jobs that are currently in progress, as well as allocating resources and scheduling. The job manager will create a unique job_id for the training job, which can be used to identify the progress of the job.

9. The job manager then sends the job_id back to the training job, which assigns it to itself.

10. The training job in turn returns the same job_id to the model training endpoint.

11. The model training endpoint creates a JSON response that contains this job_id and instructs the web server to send it back as a response to the client making the request.

12. The web server accordingly makes an HTTP response that transfers over the network and reaches the H2O client.

13. The client then creates a model object that contains this job_id, which the user can further use to track the progress of the model training or perform predictions once training is finished.

This sums up the sequence of events that take place inside the H2O server when it receives a model training request.

Now that we understand what happens to the training request, let's understand what the events that take place are when the training job created in *step 6* is training the model.

H2O runs the model training job

In H2O, the training of a model is carried out by an internal model training job that acts independently from the user's API request. The user's API request just initiates the job; the job manager does the actual execution of the job.

The following sequence diagram shows you the sequence of interactions that take place when a model training job is training a model:

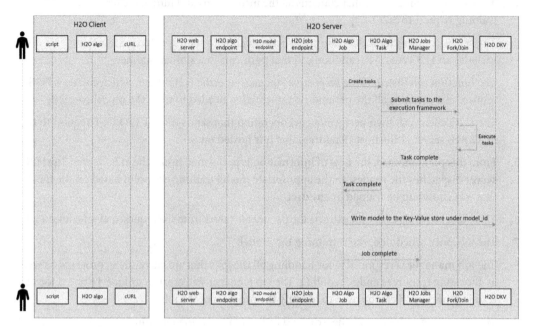

Figure 4.8 – Sequence of interactions in the model training job execution

The following set of sequences takes place during model training:

1. The model training job breaks down the model training into tasks.

2. The job then submits the tasks to the execution framework.

3. The execution framework uses the Java concurrency library `jsr166y` to perform the task in a concurrent manner using the Fork/Join processing framework.

4. Once a forked task is successfully executed, the execution library sends back the completed task results.

5. Once all the tasks are completed, the model that is trained is sent back to the model training job.

6. The model training job then stores the model object in H2O's distributed key-value storage and tags it with a unique model ID.

7. The training job then informs the job manager that model training is completed and the job manager is then free to move on to other training jobs.

Now that we understand what goes on behind the scenes when a model training job is training a model, let's move on to understand what happens when a client polls for the model training status.

Client polls for model training job completion status

As mentioned previously, the actual training of the model is processed independently from the client's training request. In this case, once a training request is sent by the client, the client is in fact unaware of the progress of the model. The client will need to constantly poll for the status of the model training job. This could be done either via manually making a request using HTTP or via certain client software features, such as progress trackers polling the H2O server for the status of the model training at regular intervals.

The following sequence diagram shows you the sequence of interactions that takes place when a client polls for the model training job completion:

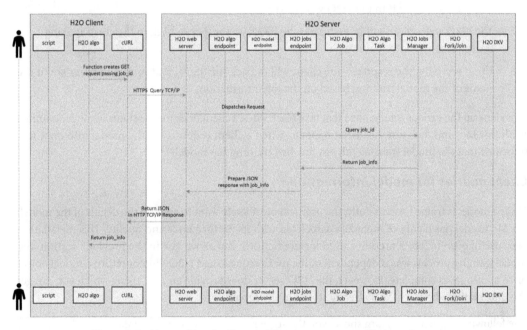

Figure 4.9 – User polling for the model status sequence of interactions

The following set of sequences takes place when the client polls for the model training job completion:

1. To get the status of the model training, the client will make a GET request, passing the `job_id` that it received as a response when it first made a request to train a model.

2. The GET request transfers over the network and to the H2O web server at the host IP address.

3. The H2O web server dispatches the request to the H2O jobs endpoint.

4. The H2O jobs endpoint will then query the jobs manager, requesting the status of the `job_id` that was passed in the GET request.

5. The job manager will return the job info of the respective `job_id` that contains information about the progress of the model training.

6. The H2O jobs endpoint will prepare a JSON response containing the job information for the `job_id` and send it to the H2O web server.

7. The H2O web server will in turn send the JSON as a response back to the client making the request.

8. Upon receiving the response, the client will unpack this JSON and update the user about the status of the model training based on the job information.

This sums up the various interactions that take place when a client polls for the status of model training. With this in mind, let's now see what happens when a client requests for the model info once it is informed that the model training job has finished training the model.

Client queries for model information

Once a model is trained successfully, the user will most likely want to analyze the details of the model. An ML model has plenty of metadata associated with its performance and quality. This metadata is very useful even before a model is used for predictions. But as we saw in the previous section, the model training process was independent of the user's request, and H2O did not return a model object once training was complete. However, the H2O server does provide an API, using which you can get the information about a model already stored in the server.

The following sequence diagram shows you the sequence of interactions that take place when a client requests information about a trained model:

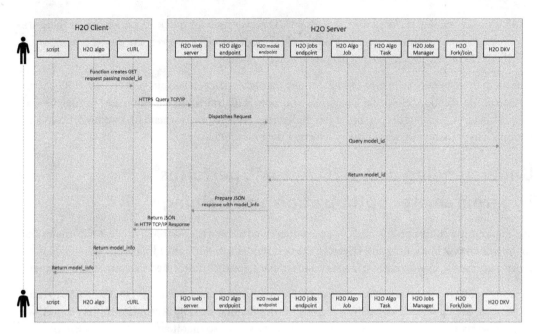

Figure 4.10 – User querying for model information

The following set of sequences takes place when the client polls for the model training job completion:

1. To get the model information, the client will make a GET request passing the unique model_id of the ML model.

2. The GET request transfers over the network and to the H2O web server at the host IP address.

3. The H2O web server dispatches the request to the H2O model endpoint.

4. All the model information is stored in H2O's distributed key-value storage when the model training job finishes training the model. The H2O model endpoint will query this distributed key-value storage with the model_id as the filter.

5. The distributed key-value storage will return all the model information for the model_id passed to it.

6. The H2O model endpoint will then prepare a JSON response containing the model information and send it to the H2O web server.

7. The H2O web server will in turn send the JSON as a response back to the client making the request.

8. Upon receiving the response, the client will extract all the model information and display it to the user.

A model, once trained, is stored directly in the H2O server itself for quick access whenever there are any prediction requests. You can download the H2O model as well; however, any model not imported into the H2O server cannot be used for predictions.

This sums up the entire sequence of interactions that takes place in various parts of the H2O client-server communication. Now that we understand how H2O trains models internally using jobs and the job manager, let's dive deeper and try to understand what happens when H2O AutoML trains and optimizes hyperparameters, eventually selecting the best model.

Understanding how H2O AutoML performs hyperparameter optimization and training

Throughout the course of this book, we have marveled at how the AutoML process automates the sophisticated task of training and selecting the best model without us needing to lift a finger. Behind every automation, however, there is a series of simple steps that is executed in a sequential manner.

Now that we have a good understanding of H2O's architecture and how to use H2O AutoML to train models, we are now ready to finally open the black box, that is, H2O AutoML. In this section, we shall understand what H2O AutoML does behind the scenes so that it automates the entire process of training and selecting the best ML models.

The answer to this question is pretty simple. H2O AutoML automates the entire ML process using **grid search hyperparameter optimization**.

Grid search hyperparameter optimization sounds very intimidating to a lot of non-experts, but the concept in itself is actually very easy to understand, provided that you know some of the basic concepts in model training, especially the importance of **hyperparameters**.

So, before we dive into grid search hyperparameter optimization, let's first come to understand what hyperparameters are.

Understanding hyperparameters

Most software engineers are aware of what parameters are: certain variables containing certain user input data, or any system-calculated data that is fed to another function or process. In ML, however, this concept is slightly complicated due to the introduction of hyperparameters. In the field of ML, there are two types of parameters. One type we call the **model parameters**, or just parameters, and the other is **hyperparameters**. Even though they have a similar name, there are some important differences between them that all software engineers should keep in mind when working in the ML space.

So, let's understand them by simple definition:

- **Model parameters**: A model parameter is a parameter value that is calculated or learned by the ML algorithm from the given dataset during model training. Some examples of basic model parameters are **mean** or **standard deviation, weights**, and **biases** of data in the dataset. These are elements that we learn from the training data when we are training the model and these are the parametric values that the ML algorithm uses to train the ML model. Model parameters are also called **internal parameters**. Model parameters are not adjustable in a given ML training scenario.

- **Hyperparameters**: Hyperparameters are configurations that are external to the model training and are not derived from the training dataset. These are parametric values that are set by the ML practitioner and are used to derive the model parameters. They are values that are heuristically discovered by the ML practitioner and input to the ML algorithm before model training begins. Some simple examples of hyperparameters are the **number of trees** in a random forest, or the **learning rate** in regression algorithms. Every type of ML algorithm will have its own required set of hyperparameters. Hyperparameters are adjustable and are often experimented with to get the optimal model in a given ML training scenario.

The aim of training an optimal model is simple:

1. You select the best combination of hyperparameters.
2. These hyperparameters generate the ideal model parameters.
3. These model parameters train a model with the lowest possible error rate.

Sounds simple enough. However, there is a catch. Hyperparameters are not intuitive in nature. One cannot simply just observe the data and decide x value for the hyperparameter will get us the best model. Finding the perfect hyperparameter is a trial-and-error process, where the aim is to find a combination that minimizes errors.

Now, the next question that arises is how you find the best hyperparameters for training a model. This is where hyperparameter optimization comes into the picture, which we will cover next.

Understanding hyperparameter optimization

Hyperparameter optimization, also known as **hyperparameter tuning**, is the process of choosing the best set of hyperparameters for a given ML algorithm to train the most optimal model. The best combination of these values minimizes a predefined **loss function** of an ML algorithm. A loss function in simple terms is a function that measures some unit of error. The loss function is different for different ML algorithms. A model with the lowest possible amount of errors among a potential combination of hyperparameter values is said to have optimal hyperparameters.

There are many approaches to implementing hyperparameter optimization. Some of the most common ones are **grid search**, **random grid search**, **Bayesian optimization**, and **gradient-based optimization**. Each is a very broad topic to cover; however, for this chapter, we shall focus on only two approaches: grid search and random grid search.

Tip

If you want to explore more about the Bayesian optimization technique for hyperparameter tuning, then feel free to do so. You can get additional information on the topic at this link: `https://arxiv.org/abs/1807.02811`. Similarly, you can get more details on gradient-based optimization at this link: `https://arxiv.org/abs/1502.03492`.

It is actually the random grid search approach that is used by H2O's AutoML for hyperparameter optimization, but you need to have an understanding of the original grid search approach to optimization in order to understand random grid search.

So, let's begin with grid search hyperparameter optimization.

Understanding grid search optimization

Let's take the example of the Iris Flower Dataset that we used in *Chapter 1, Understanding H2O AutoML Basics*. In this dataset, we are training a model that is learning from the sepal width, sepal length, petal width, and petal length to predict the classification type of the flower.

Now, the first question you are faced with is: which ML algorithm should be used to train a model? Assuming you do come up with an answer to that and choose an algorithm, the next question you will have is: which combination of hyperparameters will get me the optimal model?

Traditionally, ML practitioners would train multiple models for a given ML algorithm with different combinations of hyperparameter values. They would then compare the performance of these models and find out which hyperparameter combination trained the model with the lowest possible error rate.

The following diagram shows you how different combinations of hyperparameters train different models with varying performance:

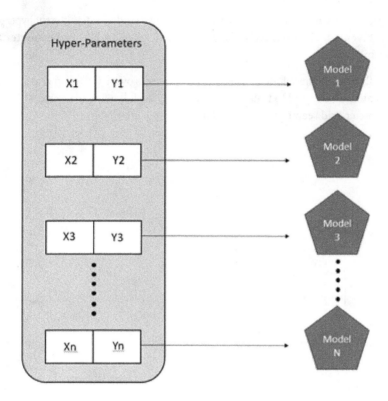

Figure 4.11 – Manual hyperparameter tuning

Let's take an example where you are training a decision tree. Its hyperparameters are the number of trees, ntrees, and the maximum depth, max_depth. If you are performing a manual search for hyperparameter optimization, then you will initially start out with values like 50, 100, 150, and 200 for ntrees and 5, 10, and 50 for max_depth, train the models, and measure their performance. When you find out which combination of those values gives you the best results, you set those values as the threshold and tweak them with smaller increments or decrements, retrain the models with these new hyperparameter values, and compare the performance again. You keep doing this until you find the best set of hyperparameter values that gives you the optimum performance.

This method, however, has a few drawbacks. Firstly, the range of values you can try out initially is limited since you can only train so many models manually. So, if you have a hyperparameter whose value can range between 1 and 10,000, then you need to make sure that you cover enough ground to not miss the ideal value by a huge margin. If you do, then you will end up constantly tweaking the value with smaller increments or decrements, spending lots of time optimizing. Secondly, as the number of hyperparameters increases and the number of possible values and combinations of values you want to use increases, it becomes tedious for the ML practitioner to manage and run optimization processes.

To manage and partially automate this process of training multiple models with different hyperparameters, grid search was invented. Grid search is also known as **Cartesian Hyperparameter Search** or **exhaustive search**.

Grid search basically maps all the values for given hyperparameters over a Cartesian grid and exhaustively searches combinations in the grid to train models. Refer to the following diagram, which shows you how a hyperparameter grid search translates to multiple models being trained:

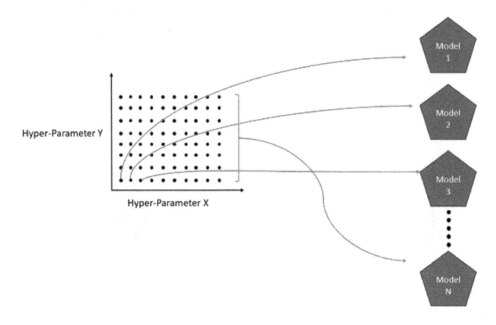

Figure 4.12 – Cartesian grid search hyperparameter tuning

In the diagram, we can see that we have a two-dimensional grid that maps the two hyperparameters. Using this Cartesian grid, we can further expand the combination of hyperparameter values to 10 values per parameter, extending our search. The grid search approach exhaustively searches across different values of the two hyperparameters. So, it will have 100 different combinations and will train 100 different models in total, all trained without needing much manual intervention.

H2O does have grid search capabilities that users can use to test out their own manually implemented grid search approach for hyperparameter optimization. When training models using grid search, H2O will map all models that it trains to the respective hyperparameter value combinations of the grid. H2O also allows you to sort all these models based on any supported model performance metrics. This sorting helps you quickly find the best-performing model based on the metric values. We shall explore more about performance metrics in *Chapter 6, Understanding H2O AutoML Leaderboard and Other Performance Metrics*.

However, despite automating and introducing a quality-of-life improvement to manual searching, there are still some drawbacks to this approach. Grid search hyperparameter optimization suffers from what is called the **curse of dimensionality**.

The curse of dimensionality was a term coined by **Richard E. Bellman** when considering problems in dynamic programming. From the point of view of ML, this concept states that as the number of hyperparameter combinations increases, the number of evaluations that the grid search will perform increases exponentially.

For example, let's say you have a hyperparameter x and you want to try out integer values 1-20. In this case, you will end up doing 20 evaluations, in other words, training 20 models. Now suppose that there is another hyperparameter y and you want to try out the values 1-20 in combination with the values for x. Your combinations will be as follows:

$$(1,1), (1,2), (1,3), (1,4), (1,5), (1,6), (1,7)....(20,20) \text{ where } (x, y)$$

Now, there are 20x20=400 combinations in total in your grid, for which your grid search optimization will end up training 400 models. Add another hyperparameter z to it and your number of combinations will skyrocket beyond management. The more hyperparameters you have, the more combinations you would try and the more combinatorial explosion will occur.

Given the time and resource sensitivity of ML, an exhaustive search is counterproductive to finding the best model. The real world has limitations, hence a random selection of hyperparameter values has often been proven to provide better results than an exhaustive grid search.

This brings us to our next approach in hyperparameter optimization, random grid search.

Understanding random grid search optimization

Random grid search replaces the previous exhaustive grid search by choosing random values from the hyperparameter search space, rather than sequentially exhausting all of them.

For example, refer to the following diagram, which shows you an example of random grid search optimization:

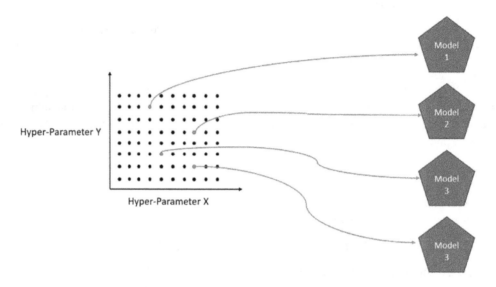

Figure 4.13 – Random grid search hyperparameter tuning

The preceding diagram is a hyperparameter space of 100 combinations of two hyperparameters, X and Y. Random grid search optimization will only choose a few at random and perform evaluations using those hyperparameter values.

The drawback of random grid search optimization is that it is a best effort approach to find the best combination of hyperparameter values with a limited number of evaluations. It may or may not find the best combination of hyperparameter values to train the optimal model, but given a large sample size, it can find the near-perfect combination to train a model with good-enough quality.

H2O library functions support random grid search optimization. It provides users with the functionality to set their own hyperparameter search grid and set a search criteria parameter to control the type and extent of the search. The search criteria can be anything, such as maximum runtime, the maximum number of models to train, or any metric. H2O will choose different hyperparameter combinations from the grid at random sequentially without repeat and will keep searching and evaluating till the search criteria are met.

H2O AutoML works slightly differently from random grid search optimization. Instead of waiting for the user to input the hyperparameter search grid, H2O has automated this part as well by already having a list of hyperparameters with all potential values for specific algorithms spaced out in the grid as default values. H2O AutoML also has provisions to include non-default values in the hyperparameter search list set by the user. H2O AutoML has predetermined values already set for algorithms; we shall explore them in the next chapter, along with understanding how different algorithms work.

Summary

In this chapter, we have come to understand the high-level architecture of H2O and what the different layers that comprise the overall architecture are. We then dived deep into the client and JVM layer of the architecture, where we understood the different components that make up the H2O software stack. Next, keeping the architecture of H2O in mind, we came to understand the flow of interactions that take place between the client and server, where we understood how exactly we command the H2O server to perform various ML activities. We also came to understand how the interactions flow down the architecture stack during model training.

Building on this knowledge, we have investigated the sequence of interactions that take place inside the H2O server during model training. We also looked into how H2O trains models using the job manager to coordinate training jobs and how H2O communicates the status of model training with the user. And, finally, we unboxed H2O AutoML and came to understand how it trains the best model automatically. We have understood the concept of hyperparameter optimization and its various approaches and how H2O automates these approaches and mitigates their drawbacks to automatically train the best model.

Now that we know the internal details of H2O AutoML and how it trains models, we are now ready to understand the various ML algorithms that H2O AutoML trains and how they manage to make predictions. In the next chapter, we shall explore these algorithms and have a better understanding of models, which will help us to justify which model would work best for a given ML problem.

5
Understanding AutoML Algorithms

All ML algorithms have a foundation in **computational statistics**. Computational statistics is the combination of statistics and computer science where computers are used to compute complex mathematics. This computation is the ML algorithm and the results that we get from it are the predictions. As engineers and scientists working in the field of ML, we are often expected to know the basic logic of ML algorithms. There are plenty of ML algorithms in the AI domain. All of them aim to solve different types of prediction problems. All of them also have their own set of pros and cons. Thus, it became the job of engineers and scientists to find the best ML algorithms that can solve a given prediction problem within the required constraints. This job, however, was eased with the invention of AutoML.

Despite AutoML taking over this huge responsibility of finding the best ML algorithm, it is still our job as engineers and scientists to verify and justify the selection of these algorithms. And to do that, a basic understanding of the ML algorithms is a must.

In this chapter, we shall explore and understand the various ML algorithms that H2O AutoML uses to train models. As mentioned previously, all ML algorithms have a heavy foundation in statistics. Statistics itself is a huge branch of mathematics and is too large to cover in a single chapter. Hence, for the sake of having a basic understanding of how ML algorithms work, we shall explore their inner workings conceptually with basic statistics rather than diving deep into the math.

> Tip
> If you are interested in gaining more knowledge in the field of statistics, then the following link should be a good place to start: `https://online.stanford.edu/courses/xfds110-introduction-statistics`.

First, we shall understand what the different types of ML algorithms are and then learn about the workings of these algorithms. We will do this by breaking them down into individual concepts, understanding them, and then building the algorithm back up to understand the big picture.

In this chapter, we are going to cover the following topics:

- Understanding the different types of ML algorithms
- Understanding the Generalized Linear Model algorithm
- Understanding the Distributed Random Forest algorithm
- Understanding the Gradient Boosting Machine algorithm
- Understanding what is Deep Learning

So, let's begin our journey by understanding the different types of ML algorithms.

Understanding the different types of ML algorithms

ML algorithms are designed to solve a specific prediction problem. These prediction problems can be anything that can provide value if predicted accurately. The differentiating factor between various prediction problems is what value is to be predicted. Is it a simple yes or no value, a range of numbers, or a specific value from a list of potential values, probabilities, or semantics of a text? The field of ML is vast enough to cover the majority, if not all, of such problems in a wide variety of ways.

So, let's start with understanding the different categories of prediction problems. They are as follows:

- **Regression**: Regression analysis is a statistical process that aims to find the relationship between independent variables, also called features, and dependent variables, also called label or response variables, and use that relationship to predict future values.

  Regression problems are problems that aim to predict certain continuous numerical values – for example, predicting the price of a car given the car's brand name, engine size, economy, and electronic features. In such a scenario, the car's brand name, engine size, economy, and electronic features are the independent variables as their presence is independent of other values, while the car price is the dependent variable whose value is dependent on the other features. Also, the price of the car is a continuous value as it can numerically range anywhere from 0 to 100 million in dollars or any other currency.

- **Classification**: Classification is a statistical process that aims to categorize the label values depending on their relationship to the features into certain classes or categories.

  Classification problems are problems that aim to predict a certain set of values – for example, predicting if a person is likely to face heart disease, depending on their cholesterol level, weight, exercise levels, heart rate, and family history. Another example would be predicting the rating of a restaurant on Google reviews that ranges from 1-5 stars, depending on the location, food, ambience, and price.

As you can see from these examples, classification problems can be either a *yes* or *no*, *true* or *false*, or *1* or *0* type of classification, or a specific set of classification values, such as those in the Google Review example, where the values can either be 1, 2, 3, 4, or 5. Thus, classification problems can be further divided, as follows:

- **Binary Classification**: In this type of classification problem, the predicted values are binary, meaning they have only two values – that is, *yes* or *no*, *true* or *false*, *1* or *0*.

- **Multiclass/Polynomial Classification**: In this type of classification problem, the predicted values are non-binary, also called polynomial, in nature, meaning they have more than two sets of values. For example, classification by age, which involves whole numbers from 1 to 100, or classification by primary colors, which can be red, yellow, or blue.

- **Clustering**: Clustering is a statistical process that aims to group or divide certain data points in such a way that data points in a single group have similar characteristics that are different from data points in other groups.

  Clustering problems are problems that aim to understand similarities within a set of values.

  For example, given a set of people who play video games with certain details, such as the hardware they use, the different games they play, and time spent playing those video games, you can categorize people by their favorite game genre. Clustering can be further divided as follows:

  - **Hard Clustering**: In this type of clustering, all data points either belong to one or another cluster; they are mutually exclusive.

  - **Soft Clustering**: In this type of clustering, rather than assigning a data point to a cluster, the probability that a data point might belong to a certain cluster is calculated. This opens the likelihood that a data point might belong to multiple clusters at the same time.

- **Association**: Association is a statistical process that aims to find the probability that if event A happened, what is the likelihood that event B will happen too? Association problems are based on association rules, which are if-then statements that show the probability of a relationship between different data points.

  The most common example of the association problem is **Market Basket Analysis**. Market Basket Analysis is a prediction problem where, given a user buys a certain product A from the market, what is the probability of the user buying product B, which is related to product A?

- **Optimization/Control**: **Control Theory**, **Optimal Control Theory**, or **Optimization Problems** is a branch of mathematics that deals with finding a certain combination of values that collectively optimize a dynamic system. **Machine Learning Control** (**MLC**) is a subfield in ML that aims to solve the Optimization Problem using ML. A good example of MLC is the implementation of ML to optimize traffic on roads using automated cars.

Now that we understand the different types of prediction problems, let's dive into understanding the different types of ML algorithms. The different types of ML algorithms are categorized as follows:

- **Supervised Learning**: Supervised learning is the ML task of mapping the relationship between the independent variables and dependent variables based on previously existing values that are labeled. Labeled data is data that contains information about which of its features are dependent and which features are independent. In supervised learning, we know which feature we want to predict and tag that feature as a label. The ML algorithm will use this information to map the relationships. Using this mapping, we predict the output for new input values. Another way of understanding this problem is that the previously existing values supervise the ML algorithm's learning task.

  Supervised learning algorithms are often used to solve regression and classification problems.

  Some examples of supervised learning algorithms are **decision trees**, **linear regression**, and **neural networks**.

- **Unsupervised learning**: As mentioned previously, supervised learning is the ML task of finding patterns and behaviors from data that is not tagged. In this case, we don't know which feature we want to predict, or the feature we want to predict may not even be a part of the dataset. Unsupervised learning helps us predict potential repeating patterns and categorize the set of data using those patterns. Another way of understanding this problem is that there are no labeled values to supervise the ML algorithm learning task; the algorithm learns the patterns and behaviors on its own.

  Unsupervised learning algorithms are often used to solve clustering and association problems.

  Some examples of unsupervised learning algorithms are **K-means clustering** and **association rule learning**.

- **Semi-supervised learning**: Semi-supervised learning falls between supervised learning and unsupervised learning. It is the ML task of performing learning on a dataset that is partially labeled. It is used in scenarios where you have a small dataset that is labeled along with a large unlabeled dataset. In real-world scenarios, labeling large amounts of data is an expensive task as it requires a lot of experimentation and contextual information that is manually interpreted, while unlabeled data is relatively cheap to acquire. Semi-supervised learning often proves efficient in this case as it is good at assuming expected label values from unlabeled datasets while working as efficiently as any supervised learning algorithm.

  Unsupervised learning algorithms are often used to solve clustering and classification problems.

  Some examples of semi-supervised learning algorithms are **generative models** and **Laplacian regularization**.

- **Reinforcement learning**: Reinforcement learning is an ML task that aims to identify the next correct logical action to take in a given environment to maximize the cumulative reward. In this type of learning, the accuracy of the prediction is calculated after the prediction is made using positive and/or negative reinforcement, which is again fed to the algorithm. This continuous learning of the environment eventually helps the algorithm find the best sequence of steps to take to maximize the reward, thus making the most accurate decision.

  Reinforcement learning is often used to solve a mix of regression, classification, and optimization problems.

  Some examples of reinforcement learning algorithms are **Monte Carlo Methods**, **Q-Learning**, and **Deep Q Network**.

The AutoML technology, despite being mature enough to be used commercially, is still in its infancy compared to the vast developments in the field of ML. AutoML may be able to train the best predictive models in the shortest time using little to no human intervention, but its potential is currently limited to only supervised learning. The following diagram summarizes the various types of ML algorithms categorized under the different ML tasks:

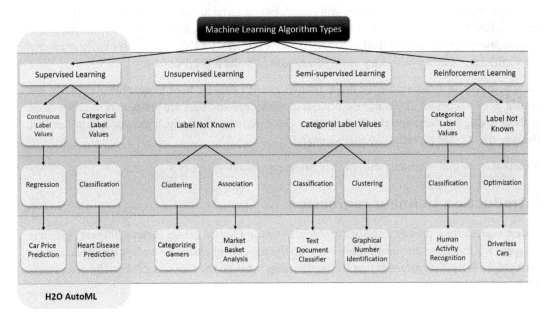

Figure 5.1 – Types of ML problems and algorithms

Similarly, H2O's AutoML also focuses on supervised learning and as such, you are often expected to have labeled data that you can feed to it.

ML algorithms that perform unsupervised learning are often quite sophisticated compared to supervised learning algorithms as there is no ground truth to measure the performance of the model. This goes against the very nature of AutoML, which is very reliant on model performance measurements to automate training and hyperparameter tuning.

So, accordingly, H2O AutoML falls in the domain of supervised ML algorithms, where it trains several supervised ML algorithms to solve regression and classification problems and ranks them based on their performance. In this chapter, we shall focus on these ML algorithms and understand their functionality so that we are well equipped to understand, select, and justify the different models that H2O AutoML trains for a given prediction problem.

With this understanding, let's start with the first ML algorithm: Generalized Linear Model.

Understanding the Generalized Linear Model algorithm

Generalized Linear Model (**GLM**), as its name suggests, is a flexible way of generalizing linear models. It was formulated by *John Nelder* and *Robert Wedderburn* as a way of combining various regression models into a single analysis with considerations given to different probability distributions. You can find their detailed paper (Nelder, J.A. and Wedderburn, R.W., 1972. *Generalized linear models. Journal of the Royal Statistical Society: Series A (General), 135(3), pp.370-384.*) at `https://rss.onlinelibrary.wiley.com/doi/abs/10.2307/2344614`.

Now, you may be wondering what linear models are. Why do we need to generalize them? What benefit does it provide? These are relevant questions indeed and they are pretty easy to understand without diving too deep into the mathematics. Once we break down the logic, you will notice that the concept of GLM is pretty easy to understand.

So, let's start by understanding the basics of linear regression.

Introduction to linear regression

Linear regression is probably one of the oldest statistical models, dating back to 200 years ago. It is an approach that maps the relationship between the dependent and independent variables linearly on a graph. What that means is that the relationship between the two variables can be completely explained by a straight line.

Consider the following example:

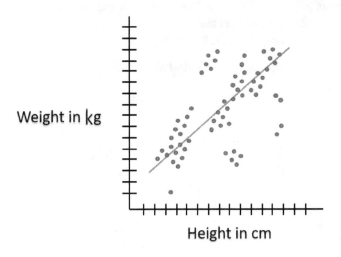

Figure 5.2 – Linear regression

This example demonstrates the relationship between two variables. The height of a person, H, is an independent variable, while the weight of a person, W, is a dependent variable. The relationship between these two variables can easily be explained by a straight red line. The taller a person is, the more likely he or she will weigh more. Easy enough to understand.

Statistically, the general equation for any straight line, also called the **linear equation**, is as follows:

$$y = b_1 x + b_0$$

Here, we have the following:

- y is a point on the Y-axis and indicates the dependent variable.
- x is a point on the X-axis and indicates the independent variable.
- b_1 is the slope of the line, also called the gradient, and indicates how steep the line is. The bigger the gradient, the steeper the line.
- b_0 is a constant that indicates the point at which the line crosses the Y-axis.

During linear regression, the machine will map all the data points of the two variables on the graph and randomly place the line on the graph. Then, it will calculate the values of y by inserting the value of x from the data points in the graph into the linear equation and comparing the result with the respective y values from the data points. After that, it will calculate the magnitude of the error between the y value it calculated and the actual y value. This difference in values is what we call a **residual**.

The following diagram should help you understand what residuals are:

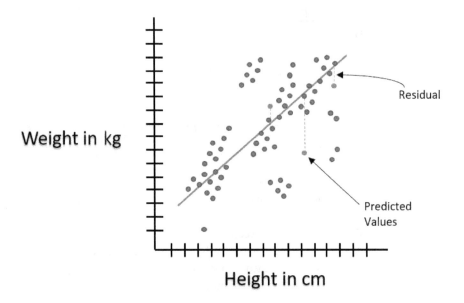

Figure 5.3 – Residuals in linear regression

The machine will do this for all the data points and make a note of all the errors. It will then try to tweak the line by changing the values of b_1 and b_0, meaning changing the angle and position of the line on the graph, and repeating the process. It will do this until it minimizes the error.

The values of b_1 and b_0 that generate the least amount of error are the most accurate linear relationship between the two variables. The equation with these values for b1 and b_0 is the linear model.

Now, say you want to predict how much a person would weigh if they were 180 cm tall. Then, you use this same linear model equation with the b_1 and b_0 values, set x to 180, and calculate y, which will be the expected weight.

Congratulations, you just performed ML in your mind without any computers and made predictions too! Actual ML works the same way, albeit with added complexities from complex algorithms. Linear regression doesn't need to be restricted to just two variables – it can also work on multiple variables where there's more than one independent variable. Such linear regression is called multiple or curvilinear regression. The equation of such a linear regression expands as follows:

$$y = b_1 x_1 + b_2 x_2 + b_3 x_3 + \cdots + b_0$$

In this equation, the additional variables – x_1, x_2, x_3, and so on – are added with their own coefficients – b_1, b_2, and b_3, respectively.

Feel free to explore these algorithms and the mathematics behind them if you are interested in the inner workings of linear regression.

Understanding the assumptions of linear regression

Linear regression, when training a model on a given dataset, works on certain assumptions about the data. One of these assumptions is the **normality of errors**.

Before we understand what the normality of errors is, let's quickly understand the concept of the **probability density function**. This is a mathematical expression that defines the probability distribution of discrete values – in other words, it is a mathematical expression that shows the probability of a sample value occurring from a given sample space. To understand this, refer to the following diagram:

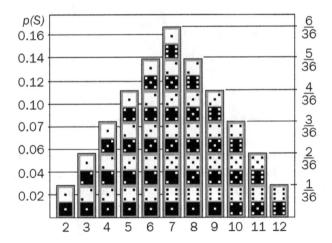

Figure 5.4 – Probability distribution of values for two dice

The preceding diagram shows the distribution of probabilities of all the values that can occur when a pair of six-sided dice are thrown fairly and independently. There are different kinds of distributions. Some examples of commonly occurring distributions are as follows:

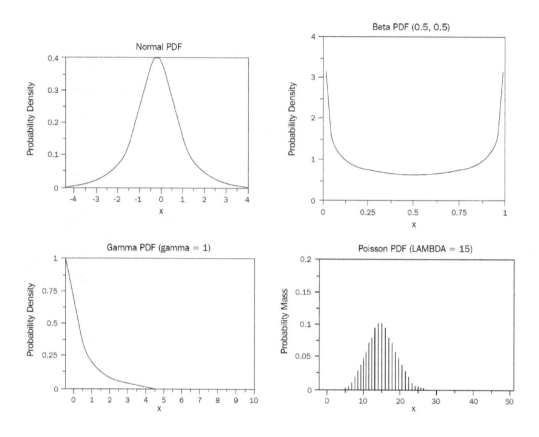

Figure 5.5 – Different types of distribution

The normality of errors states that the residuals of the data must be normally distributed. A **normal distribution**, also called **Gaussian distribution**, is a probability density function that is symmetrical about the mean, where the values closest to the mean occur frequently, while those far from the mean rarely occur. The following diagram shows a normal distribution:

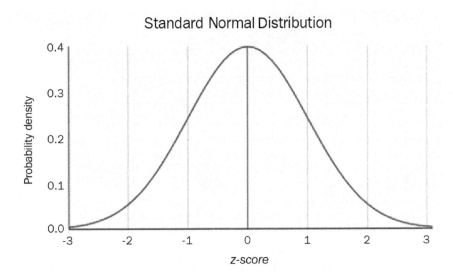

Figure 5.6 – Normal distribution

Linear regression expects the residuals that get calculated to fall within a normal distribution. In our previous example of the expected weight for a height, there is bound to be some error between the predicted weight and the actual weight of a person with a certain height. However, the residuals or errors from the prediction will most likely fall within a normal distribution as there cannot be too many occurrences of people with an extreme difference between the expected weight and the predicted weight.

Consider a scenario of people claiming health insurance payouts. The following diagram shows a sample of the linear regression graph for that dataset:

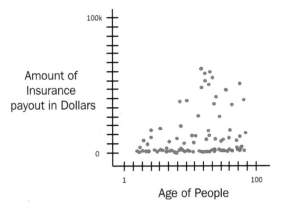

Figure 5.7 – Health insurance payout

In the preceding diagram, you can see that the majority of people from various age groups did not claim health insurance. Some of them did and the cost of claims varied a lot. Some had minor issues costing *little*, while some had serious injuries and had to go through expensive surgeries.

If you plot a linear regression line through this dataset, it will look as follows:

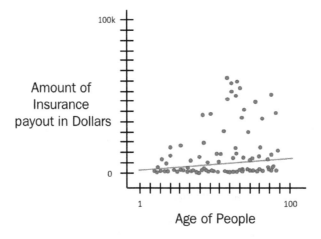

Figure 5.8 – Linear regression on health insurance payouts

But now, if you calculate the residual errors from the expected and predicted value for all the data points, then the probability distribution of these residuals will not fall into a normal distribution. It will look as follows:

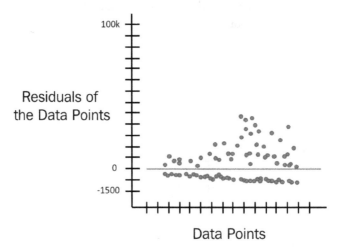

Figure 5.9 – Residual distribution of health insurance payouts

This is an inaccurate model as the expected value and the predicted values are not even close enough to round off or correct. So, what do you do for such a scenario, where the normality of errors assumption fails for the dataset? What if the distribution of the residuals is, say, Poisson instead of normal? How will the machine correct that?

Well, the answer to this is that the distribution of residuals depends on the distribution of the dataset itself. If the values of the dependent variables are normally distributed, then the distribution of the residuals will also be normal. So, once we have identified which probability density function fits the dataset, we can use that function to train our linear model.

Depending on this function, there are specialized linear regression methods for every probability density function. If your distribution is **Poisson**, then you can use **Poisson regression**. If your data distribution is negative **binomial**, then you can use **negative binomial regression**.

Working with a Generalized Linear Model

Now that we have covered the basics, let's focus on understanding what GLM is. GLM is a way of pointing to all the regression methods that are specific to the type of probability distribution of the data. Technically, all the regression models are GLM, including our ordinary simple linear model. GLM just encapsulates them together and trains the appropriate regression model based on the probability distribution function.

The way GLM works is by using something called a link function in conjunction with a systematic component and the random variable.

These are three components of GLM:

- **Systematic component**: Going back to the multi-variate linear equation, we have the following:

$$y = b_1 x_1 + b_2 x_2 + b_3 x_3 + \cdots + b_0$$

  Here, $b_1 x_1 + b_2 x_2 + b_3 x_3 + \ldots + b_0$ is the systematic component. This is the function that links our data, also called predictors, with our predictions.

- **Random Component**: This component refers to the probability distribution of the response variable. This will be whether the response variable is normally distributed or binomially distributed or any other form of distribution.

- **Link function**: A link function is a function that maps the non-linear relationship of data to a linear one. In other words, it bends the line of linear regression to represent the relationship of non-linear data more accurately. It is a link between the random and the systematic components. We can explain the equation with a link function mathematically as $Y = f_n(b_1 x_1 + b_2 x_2 + b_3 x_3 + \ldots + b_0)$, where f_n is the link function that changes as per the distribution of the response variable.

The link function is different for different distributions. The following table shows the different link functions for different distributions:

Distribution Type	Link Function	Name of Algorithm
Normal	$b_0 + b_1 x$	Linear model
Binomial	$\dfrac{e^{b_0 + b_1 x}}{1 + e^{b_0 + b_1 x}}$	Logistic regression
Poisson	$e^{b_0 + b_1 x}$	Poisson regression
Gamma	$\dfrac{1}{b_0 + b_1 x}$	Gamma regression

Figure 5.10 – Link functions for different distribution types

When training GLM models, you have the option of selecting the value for the family hyperparameter. The family option specifies the probability distribution of your response column and the GLM training algorithm uses the appropriate link function during training.

The values for the family hyperparameter are as follows:

- **gaussian**: You should select this option if the response is a real integer number.
- **binomial**: You should select this option if the response is categorical with two classes or binaries that could be either enums or integers.
- **fractionalbinomial**: You should select this option if the response is numeric between 0 and 1.
- **ordinal**: You should select this option if the response is a categorical response with three or more classes.
- **quasibinomial**: You should select this option if the response is numeric.
- **multinomial**: You should select this option if the response is a categorical response with three or more classes that are of enum types.
- **poisson**: You should select this option if the response is numeric and contains non-negative integers.
- **gamma**: You should select this option if the response is numeric and continuous and contains positive real integers.
- **tweedie**: You should select this option if the response is numeric and contains continuous real values and non-negative values.
- **negativebinomial**: You should select this option if the response is numeric and contains a non-negative integer.
- **AUTO**: This determines the family automatically for the user.

As you may have guessed, H2O's AutoML selects AUTO as the family type when training GLM models. The AutoML process handles this case of selecting the correct distribution family by understanding the distribution of the response variable in the dataset and applying the correct link function to train the GLM model.

Congratulations, we have just looked into how the GLM algorithm works! GLM is a very powerful and flexible algorithm and H2O AutoML expertly configures its training so that it trains the most accurate and high-performance GLM model.

Now, let's move on to the next ML algorithm that H2O trains: **Distributed Random Forest (DRF)**.

Understanding the Distributed Random Forest algorithm

DRF, simply called **Random Forest**, is a very powerful supervised learning technique often used for classification and regression. The foundation of the DRF learning technique is based on **decision trees**, where a large number of decision trees are randomly created and used for predictions and their results are combined to get the final output. This randomness is used to minimize the bias and variance of all the individual decision trees. All the decision trees are collectively combined and called a forest, hence the name Random Forest.

To get a deeper conceptual understanding of DRF, we need to understand the basic building block of DRF – that is, a decision tree.

Introduction to decision trees

In very simple terms, a decision tree is just a set of *IF* conditions that either return a yes or a no answer based on data passed to it. The following diagram shows a simple example of a decision tree:

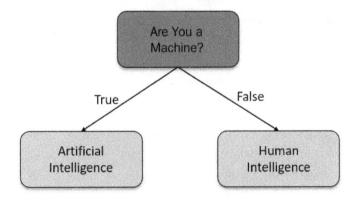

Figure 5.11 – Simple decision tree

The preceding diagram shows a basic decision tree. A decision tree consists of the following components:

- **Nodes**: Nodes are basically *IF* conditions that split the decision tree based on whether the condition was met or not.

- **Root Node**: The node on the top of the decision tree is called the root node.

- **Leaf Node**: The nodes of the decision tree that do not branch out further are called leaf nodes, or simply leaves. The condition, in this case, is if the value of the data that's passed to it is numeric, then the answer is the data is a number; if the data that's passed to it is not numeric, then the answer will be the data is non-numeric. This is simple enough to understand.

As seen in *Figure 5.11*, the decision tree is based on a simple true or false question. Decision trees can also be based on mathematical conditions on numeric data. The following example shows a decision tree on numeric conditions:

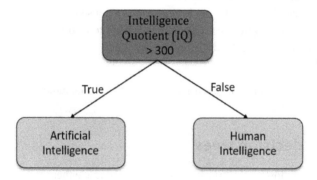

Figure 5.12 – Numerical decision tree

In this example, the root node computes whether the IQ number is greater than 300 and decides if it is artificial intelligence or human intelligence.

Decision trees can be combined as well. They can form a complex set of decision-making conditions that rely on the results of previous decisions. Refer to the following example for a complex decision tree:

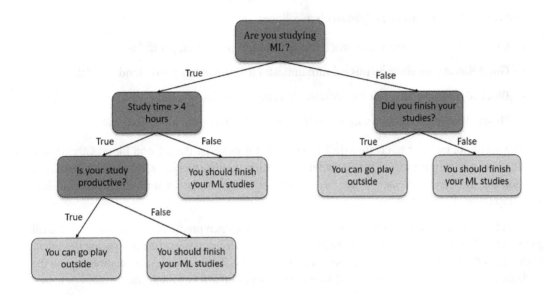

Figure 5.13 – Complex decision tree

In the preceding example, we are trying to calculate if *you can go play outside* or *finish your ML studies*. This decision tree combines numeric data as well as the classification of data. When making predictions, the decision tree will start at the top and work its way down, making decisions on whether the data satisfies the conditions or not. The leaf nodes are the final potential results of the decision tree.

With this knowledge in mind, let's create a decision tree on a sample dataset. Refer to the following table for the sample dataset:

Chest Pain	Good Blood Circulation	Blocked Arteries	Heart Disease
False	False	False	False
True	True	True	True
True	True	False	False
True	False	N/A	True
False	False	True	True

Figure 5.14 – Sample dataset for creating a decision tree

The content of the aforementioned dataset is as follows:

- **Chest Pain**: This column indicates if a patient suffers from chest pain.

- **Good Blood Circulation**: This column indicates if a patient has good blood circulation.

- **Blocked Arteries**: This column indicates if a patient has any blocked arteries.

- **Heart Disease**: This column indicates if the patient suffers from heart disease.

For this scenario, we want to create a decision tree that uses Chest Pain, Good Blood Circulation, and Blocked Arteries features to predict whether a patient has heart disease. Now, when forming a decision tree, the first thing that we need to do is find the root node. So, what feature should we place at the top of the decision tree?

We start by looking at how the Chest Pain feature alone fairs when predicting heart disease. We shall go through all the values in the dataset and map them to this decision tree while comparing the values in the Chest Pain column with those of heart disease. We shall keep track of these relationships in the decision tree. The decision tree for Chest Pain as the root node will be as follows:

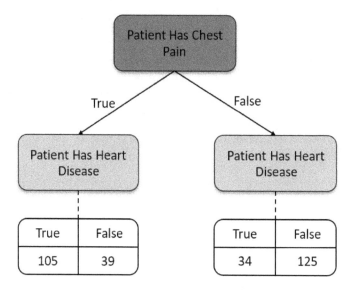

Figure 5.15 – Decision tree for the Chest Pain feature

Now, we do this for all the other features in the dataset. We create a decision tree for Good Blood Circulation and see how it fairs alone when making predictions for Heart Disease and keep a track of the comparison, repeating the same process for the Blocked Arteries status as well. If there are any missing values in the dataset, then we skip them. Ideally, you should not work with datasets that have missing values. We can use the techniques we learned about in *Chapter 3, Understanding Data Processing*, where we impute and handle missing dataset values.

Refer to the following diagram, which shows the two decision trees that were created – one for **Patient Has Blocked Arteries** and another for **Patient Has Good Blood Circulation**:

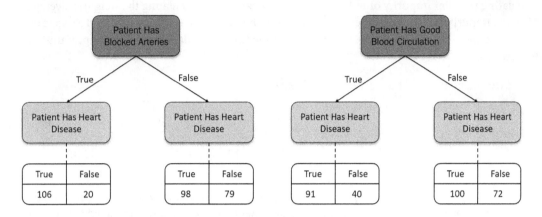

Figure 5.16 – Decision tree for the Blocked Arteries and Good Blood Circulation features

Now that we have created a decision tree for all the features in the dataset, we can compare their results to find the pure feature. In the context of decision trees, a feature is said to be 100% impure when a node is split evenly, 50/50, and 100% pure when all of its data belongs to a single class. In our scenario, we don't have any feature that is 100% pure. All of our features are impure to some degree. So, we need to find some way of finding the feature that is the purest. For that, we need a metric that can measure the purity of a decision tree.

There are plenty of ways by which data scientists and engineers measure purity. The most common metric to measure impurity in decision trees is **Gini Impurity**. Gini Impurity is the measure of the likelihood that a new random instance of data will be incorrectly classified during classification.

Gini Impurity is calculated as follows:

$$Gini\ Impurity\ =\ 1-\ (p_1^2\ +\ p_2^2\ +\ p_3^2\ +\ p_4^2\ +\ p_5^2\ ...)$$

Here, p_1, p_2, p_3, p_4 ... are the probabilities of the various classifications for Heart Disease. In our scenario, we only have two classifications – either a yes or a no. Thus, for our scenario, the measure of impurity is as follows:

$$Gini\ Impurity\ =\ 1-\ ((Probability\ of\ Yes)^2\ +\ (Probability\ of\ No)^2)$$

So, let's calculate the Gini Impurity of all the decision trees we just created so that we can find the feature that is the purest. Gini Impurity for a decision tree with multiple leaf nodes is calculated by calculating the Gini Impurity of individual leaf nodes and then calculating the weighted average of all the impurity values to get the Gini Impurity of the decision tree as a whole. So, let's start by calculating the Gini Impurity of the left leaf node of the Chest Pain decision tree and repeat this for the right leaf node:

$$Gini\ Impurity\ (Left\ Leaf\ node\ of\ Chest\ Pain)$$
$$= 1 - ((105 / (105 + 39))^2 + (39 / (105 + 39))^2) = 0.395$$

$$Gini\ Impurity\ (Right\ Leaf\ node\ of\ Chest\ Pain)$$
$$= 1 - ((34 / (34 + 125))^2 + (125 / (34 + 125))^2) = 0.336$$

The reason why we calculate the weighted average of the Gini Impurities is because the representation of the data is not equally divided between the two branches of the decision tree. The weighted average helps us offset this unequal distribution of the data values. Thus, we can calculate the Gini Impurity of the whole Chest Pain decision tree as follows:

Gini Impurity (Chest Pain) = weighted average of the Gini Impurities of the leaf nodes

Gini Impurity (Chest Pain) =

(Total number of data inputs in the left leaf node / total number of rows) x Gini Impurity of the left leaf node

+

(Total number of data inputs in the right leaf node / total number of rows) x Gini Impurity of the right leaf node

= (144 / (144 + 159)) x 0.395 + (159 / (144 + 159)) x 0.364

= 0.364

The Gini Impurity of the Chest Pain decision tree is 0.364.

We repeat this process for all the other feature decision trees as well. We should get the following results:

- The Gini Impurity of the Chest Pain decision tree is 0.364
- The Gini Impurity of the Good Blood Circulation tree is 0.360
- The Gini Impurity of the Blocked Arteries tree is 0.381

Comparing these values, we can infer that the Gini Impurity of the Good Blood Circulation feature has the lowest Gini Impurity, making it the purest feature in the dataset. So, we will use it as the root of our decision tree.

Referring to *Figures 5.12* and *5.13*, when we divided the patients by using the Good Blood Circulation feature, we were left with an impure distribution of the results on the left and right leaf nodes. So, each leaf node had a mix of results that showed with and without Heart Disease. Now, we need to figure out a way to separate the mix of results from the Good Blood Circulation feature using the remaining features – Chest Pain and Blocked Arteries.

So, just as how we did previously, we shall use these mixed results and separate them using the other features and calculate the Gini Impurity value of those features. We shall choose the feature that is the purest and replace it at the given node for further classification.

We shall repeat this process for the right branch as well. So, to simplify the selection of the decision tree nodes, we must do the following:

- Calculate the Gini Impurity score of all the remaining features for that node using the mixed results.

- Choose the one with the lowest impurity and replace it with the node.

- Repeat the same process further down the decision tree with the remaining features.

- Continue replacing the nodes, so long as the classification lowers the Gini Impurity; otherwise, leave it as a leaf node.

So, your final decision tree will be as follows:

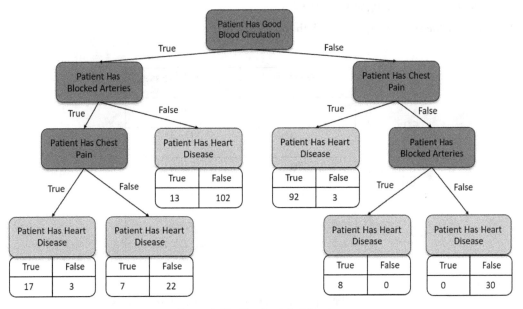

Figure 5.17 – The final decision tree

This decision tree is good for classification with true or false values. What if you had numerical data instead?

Creating decision trees with numerical data is very easy and has almost the same steps as we do for true/false data. Consider Weight as a new feature; the data for the **Weight** column is as follows:

Chest Pain	Good Blood Circulation	Blocked Arteries	Weight	Heart Disease
False	False	False	90	False
True	True	True	70	True
True	True	False	88	False
True	False	N/A	50	True
False	False	True	80	True

Figure 5.18 – Dataset with a new feature, Weight, in kilograms

For this scenario, we must follow these steps:

1. Sort the data in ascending order. In our scenario, we shall sort the rows of the dataset with the Weight column from highest to lowest.

2. Calculate the average weights for all the adjacent rows:

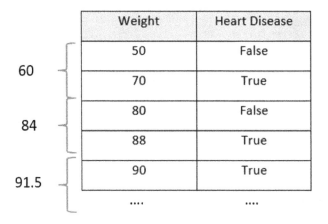

Figure 5.19 – Calculating the average of the subsequent row values

3.   Calculate the Gini Impurity of all the averages we calculated:

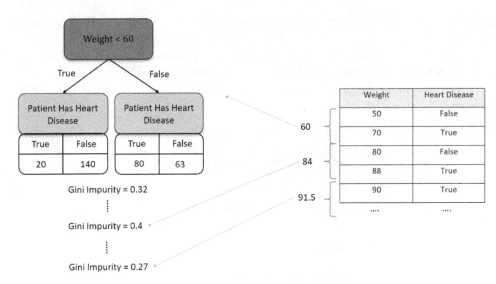

Figure 5.20 – Calculating the Gini Impurity of all the averages

4.   Identify and select the average feature value that gives us the least Gini Impurity value.

5.   Use the selected feature value as a decision node in the decision tree.

Making predictions using decision trees is very easy. You will have data with values for chest pain, blocked arteries, good circulation, and weight and you will feed it to the decision tree model. The model will filter the values down the decision tree while calculating the node conditions and eventually arriving at the leaf node with the prediction value.

Congratulations – you have just understood the concept of decision trees! Despite decision trees being easy to understand and implement, they are not that good at solving real-life ML problems.

There are certain drawbacks to using decision trees:

- Decision trees are very unstable. Any minor changes in the dataset can drastically alter the performance of the model and prediction results.

- They are inaccurate.

- They can get very complex for large datasets with a large number of features. Imagine a dataset with 1,000 features – the decision tree for this dataset will have a tree whose depth will be very large and its computation will be very resource-intensive.

To mitigate all these drawbacks, the Random Forest algorithm was developed, which builds on top of decision trees. With this knowledge, let's move on to the next concept: Random Forest.

Introduction to Random Forest

Random Forest, also called **Random Decision Forest**, is an ML method that builds a large number of decision trees during learning and groups, or ensembles, the results of the individual decision trees to make predictions. Random Forest is used to solve both classification and regression problems. For classification problems, the class value predicted by the majority of the decision trees is the predicted value. For regression problems, the mean or average prediction of the individual trees is calculated and returned as the prediction value.

The Random Forest algorithm follows these steps for learning during training:

1. Create a bootstrapped dataset from the original dataset.
2. Randomly select a subset of the data features.
3. Start creating a decision tree using the selected subset of features, where the feature that splits the data the best is chosen as the root node.
4. Select a random subset of the other remaining features to further split the decision tree.

Let's understand this concept of Random Forest by creating one.

We shall use the same dataset that we used to make our complex decision tree in *Figure 5.17*. The dataset is the same one we used to make our decision trees. To create a Random Forest, we need to create a bootstrapped version of the dataset.

A bootstrapped dataset is a dataset that is created from the original dataset by randomly selecting rows from the dataset. The bootstrapped dataset is the same size as the original dataset and can also contain duplicate rows from the original dataset. There are plenty of inbuilt functions for creating a bootstrapped dataset and you can use any of them to create one.

Consider the following bootstrapped dataset:

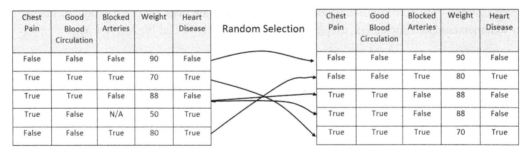

Figure 5.21 – Bootstrapping dataset

The next step is to create a decision tree from the bootstrapped dataset but using only a subset of the feature columns at each step. So, selecting all the features to be considered for the decision tree only lets you go with Good Blood Circulation and Blocked Arteries as features for the decision tree.

We shall follow the same purity identification criteria to determine the root of the node. Let's assume that for our experiment, Good Blood Circulation is the purest. Setting that as the root node, we shall now consider the remaining features to fill the next level of decision nodes. Just like we did previously, we shall randomly select two features from the remaining features and decide which feature should fit in the next decision node. We will build the tree as usual while considering the random subset of remaining variables at each step.

Here is the tree we just made:

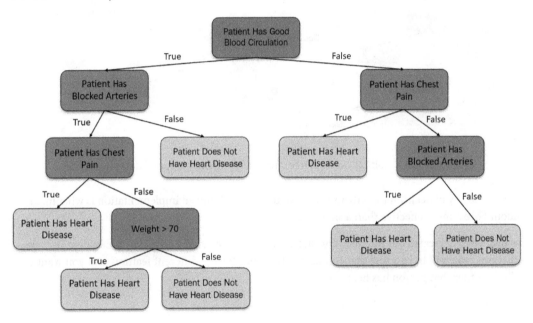

Figure 5.22 – First decision tree from the bootstrapped dataset

Now, we repeat the same process while creating multiple decision trees and bootstrapping and selecting features from random trees. An ideal Random Forest will create hundreds of decision trees, as follows:

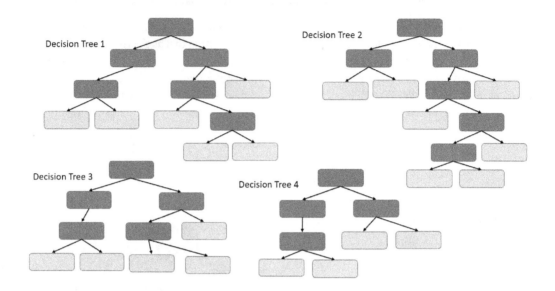

Figure 5.23 – Multiple decision trees from different bootstrapped datasets

This large variety of decision trees that were created with randomized implementation is what makes Random Forest more effective than a single decision tree.

Now that we have created our Random Forest, let's see how we can use it to make predictions. To make predictions, you will have a row that contains data values for the different features and you want to predict whether that person has heart disease.

You will pass this data down an individual decision tree in the Random Forest:

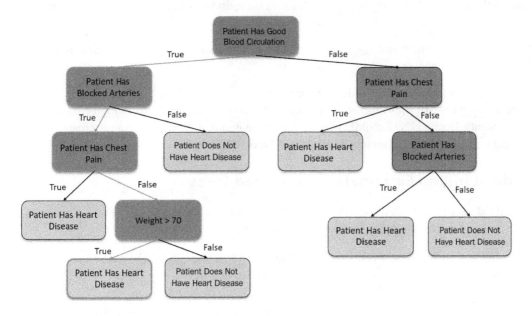

Figure 5.24 – Predictions from the first decision tree in the Random Forest

The decision tree will predict the results based on its structure. We shall keep a track of the prediction made by this tree and continue passing the data down to the other trees, noting their predictions as well:

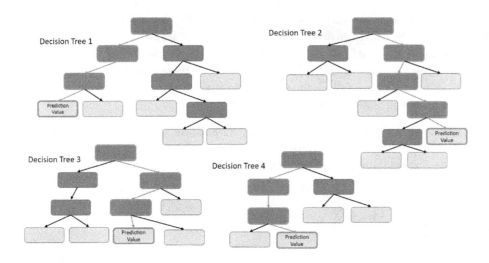

Figure 5.25 – Predictions from the other individual trees in the Random Forest

Once we get predictions from all the individual trees, we can find out which value got the most votes from all the decision trees. The prediction value with the most votes concludes the prediction for the Random Forest.

Bootstrapping the dataset and aggregating the prediction values of all the decision trees to make a decision is called **bagging**.

Congratulations – you have just understood the concept of Random Forest! Random Forest, despite being a very good ML algorithm with low bias and variance, still suffers from high computation requirements. Hence, H2O AutoML, instead of training Random Forest, trains an alternate version of Random Forest called **Extremely Randomized Trees (XRT)**.

Understanding Extremely Randomized Trees

The **XRT** algorithm, also called **ExtraTrees**, is just like the ordinary Random Forest algorithm. However, there are two key differences between Random Forest and XRT, as follows:

- In Random Forest, we use a bootstrapped dataset to train the individual decision trees. In XRT, we use the whole dataset to train the individual decision trees.

- In Random Forest, the decision nodes are split based on certain selection criteria such as the impurity metric or error rate when building the individual decision tree. In XRT, this process is completely randomized and the one with the best results is chosen.

Let's consider the same example we used to understand Random Forest to understand XRT. We have a dataset, as shown in *Figure 5.17*. Instead of bootstrapping the data, as we did in *Figure 5.20*, we shall use the dataset as-is.

Then, we start creating our decision trees by randomly selecting a subset of the features. In Random Forest, we used the purity criteria to decide which feature should be set as the root node of the decision tree. However, for XRT, we shall set the root node as well as the decision nodes of the decision tree randomly. Similarly, we shall create multiple decision trees like these with all the features randomly selected. This added randomness allows the algorithm to further reduce the variance of the model, at the expense of a slight increase in bias.

Congratulations! We have just investigated how the XRT algorithm uses an extremely randomized forest of decision trees to make accurate regressions and classification predictions. Now, let's understand how the GBM algorithm trains a classification model to classify data.

Understanding the Gradient Boosting Machine algorithm

Gradient Boosting Machine (GBM) is a forward learning ensemble ML algorithm that works on both classification as well as regression. The GBM model is an ensemble model just like the DRF algorithm in the sense that the GBM model, as a whole, is a combination of multiple weak learner models whose results are aggregated and presented as a GBM prediction. GBM works similarly to DRF in that it consists of multiple decision trees that are built in a sequence that sequentially minimizes the error.

GBM can be used to predict continuous numerical values, as well as to classify data. If GBM is used to predict continuous numerical values, we say that we are using GBM for regression. If we are using GBM to classify data, then we say we are using GBM for classification.

The GBM algorithm has a foundation on decision trees, just like DRF. However, how the decision trees are built is different compared to DRF.

Let's try to understand how the GBM algorithm works for regression.

Building a Gradient Boosting Machine

We shall use the following sample dataset and understand how GBM works as we conceptually build the model. The following table contains a sample of the dataset:

Height	Gender	Age	Weight
170	M	45	50
169	F	26	67
180	M	58	55
185	F	66	45
177	M	45	86
174	M	36	90
182	M	75	77
165	F	43	56
160	F	34	66

Figure 5.26 – Sample dataset for GBM

This is an arbitrary dataset that we are using just for the sake of understanding how GBM will build its ML model. The contents of the dataset are as follows:

- **Height**: This column indicates the height of the person in centimeters.

- **Gender**: This column indicates the gender of the person.

- **Age**: This column indicates the age of the person.

- **Weight**: This column indicates the weight of the person.

GBM, unlike DRF, starts creating its weak learner decision trees from leaf nodes instead of root nodes. The very first leaf node that it will create will be the average of all the values of the response variable. So, accordingly, the GBM algorithm will create the leaf node, as shown in the following diagram:

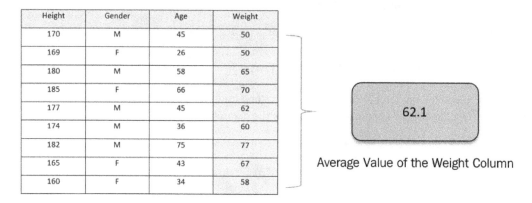

Figure 5.27 – Calculating the leaf node using the column average

This leaf node alone can also be considered a decision tree. It acts like a prediction model that only predicts a constant value for any kind of input data. In this case, it's the average value that we get from the response column. This is, as we expect, an incorrect way of making any predictions, but it is just the first step for GBM.

The next thing GBM will do is create another decision tree based on the errors it observed from its initial leaf node predictions on the dataset. An error, as we discussed previously, is nothing but the difference between the observed weight and the predicted weight and is also called the residual. However, these residuals are different from the actual residuals that we will get from the complete GBM model. The residuals that we get from the weak learner decision trees of GBM are called **pseudo-residuals**, while those of the GBM model are the actual residuals.

So, as mentioned previously, the GBM algorithm will calculate the pseudo-residuals of the first leaf node for all the data values in the dataset and create a special column that keeps track of these pseudo-residual values.

Refer to the following diagram for a better understanding:

Height	Gender	Age	Weight	Pseudo-residual 1
170	M	45	50	-12.1
169	F	26	50	-12.1
180	M	58	65	2.9
185	F	66	70	7.9
177	M	45	62	-0.1
174	M	36	60	-2.1
182	M	75	77	14.9
165	F	43	67	4.9
160	F	34	58	-4.1

Figure 5.28 – Dataset with pseudo-residuals

Using these pseudo-residual values, the GBM algorithm then builds a decision tree using all the remaining features – that is, Height, Favorite Color, and Gender. The decision tree will look as follows:

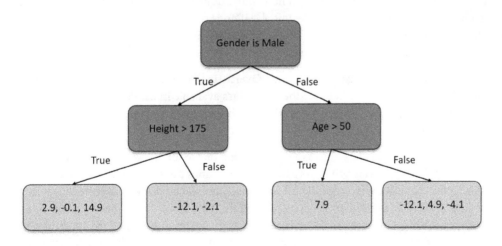

Figure 5.29 – Decision tree using pseudo-residual values

As you can see, this decision tree only has four leaf nodes, while the pseudo-residual values in the algorithm generated from the first tree are way more than four. This is because the GBM algorithm restricts the size of the decision trees it makes. For this scenario, we are only using four leaf nodes. Data scientists can control the size of the trees by passing the right hyperparameters when configuring the GBM algorithm. Ideally, for large datasets, you often use 8 to 32 leaf nodes.

Due to the restriction of the leaf nodes in the decision trees, the decision tree ends up with multiple pseudo-residual values in the same leaf nodes. So, the GBM algorithm replaces them with their average to get one concrete number for a single leaf node. Accordingly, after calculating the averages, we will end up with a decision tree that looks as follows:

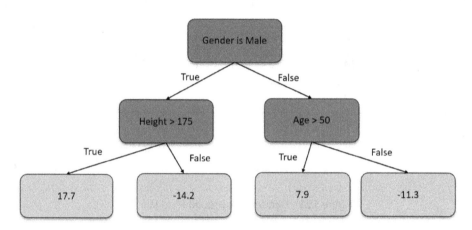

Figure 5.30 – Decision tree using averaged pseudo-residual values

Now, the algorithm combines the original leaf node with this new decision tree to make predictions on it. So, now, we have a value of 71.2 from the initial leaf node prediction. Then, after running the data down the decision tree, we get 16.8. So, the predicted weight is the summation of both the predictions, which is 88. This is also the observed weight.

This is not correct as this is a case of overfitting. **Overfitting** is a modeling error where the model function is too fine-tuned to predict only the data values available in the dataset and not any other values outside the dataset. As a result, the model becomes useless for predicting any values that fall outside of the dataset.

So, to correct this, the GBM algorithm assigns a learning rate to all the weak learner decision trees that it trains. The **learning rate** is a hyperparameter that tunes the rate at which the model learns new information that can override the old information. The value of the learning rate ranges from 0 to 1. By adding this learning rate to the predictions from the decision trees, the algorithm controls the influence of the decision tree's predictions and slowly moves toward minimizing the error step by step. For our example, let's assume that the learning rate is 0.1. So, accordingly, the predicted weight can be calculated as follows:

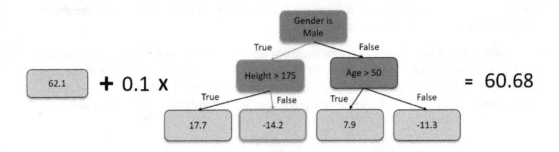

Figure 5.31 – Calculating the predicted weight

Thus, the algorithm will plug in the learning rate for the predictions made by the decision tree and then calculate the predicted weight. Now, the predicted weight will be *62.1 + (0.1 x -14.2) = 60.68.*

60.68 is not a very good prediction but it is still a better prediction than 62.68, which is what the initial leaf node predicted. The incremental steps to minimize the errors are the right way to maintain low variance in predictions. A correct balance of learning rate is also important as too high a learning rate will offshoot the correction in the opposite direction, while too low a learning rate will lead to long computation time as the algorithm will take very small correction steps to reach the minimum error.

To further correct the prediction value and minimize the error, the GBM algorithm will create another decision tree. For this, it will calculate the new pseudo-residual values from predictions made with the leaf node and the first decision tree and use these values to build the second decision tree.

The following diagram shows how new pseudo-residual values are calculated:

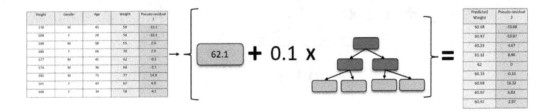

Figure 5.32 – Calculating new pseudo-residual values

You will notice that the new pseudo-residual values that were generated are a lot closer to the actual values compared to the first pseudo-residual values. This indicates that the GBM model is slowly minimizing errors and improving its accuracy.

Moving on with the second decision tree, the algorithm uses the new pseudo residual values to create the second decision tree. Once created, it aggregates the tree, along with the learning rate, to the already existing leaf node and the first decision tree.

The decision trees can be different each time the GBM algorithm creates one. However, the learning rate stays common for all the trees. So, now, the prediction values will be the summation of the three components – the initial leaf node prediction value, the scaled value of the first decision tree prediction, and the scaled value of the second decision tree prediction. So, the prediction values will be as follows:

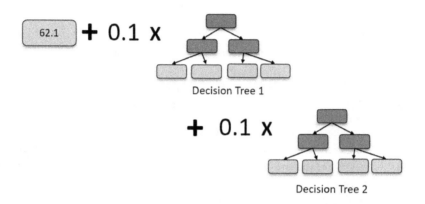

Figure 5.33 – GBM model with a second boosted decision tree

The GBM algorithm will repeat the same process, creating decision trees up to the specified number of trees or until adding decision trees stops improving the predictions. So, eventually, the GBM model will look as follows:

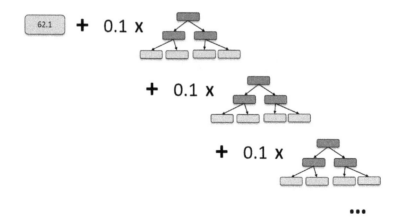

Figure 5.34 – Complete GBM model

Congratulations! we have just explored how the GBM algorithm uses an ensemble of weak decision tree learners to make accurate regressions predictions.

Another algorithm that H2O AutoML uses, which builds on top of GBM, is the XGBoost algorithm. XGBoost stands for Extreme Gradient Boosting and implements a process called boosting that sometimes helps in training better-performing models. It is one of the most widely used ML algorithms in Kaggle competitions and has proven to be an amazing ML algorithm that can be used for both classification and regression. The mathematics behind how XGBoost works can be slightly difficult for users not well versed with statistics. However, it is highly recommended that you take the time and learn more about this algorithm. You can find more information about how H2O performs XGBoost training at `https://docs.h2o.ai/h2o/latest-stable/h2o-docs/data-science/xgboost.html`.

> **Tip**
> Ensemble ML is a method of combining multiple ML models to obtain better prediction results compared to the performance of the models individually – just like how a combination of decision trees creates the Random Forest algorithm using bagging and how the GBM algorithm uses a combination of weak learners to minimize errors. Ensemble models take things one step further by finding the best combinations of prediction algorithms and using their combined performance to train a meta-learner that provides improved performance. This is done using a process called stacking. You can find more information about how H2O trains these stacked ensemble models at `https://docs.h2o.ai/h2o/latest-stable/h2o-docs/data-science/stacked-ensembles.html#stacked-ensembles`.

Now, let's learn how deep learning works and understand neural networks.

Understanding what is Deep Learning

Deep Learning (**DL**) is a branch of ML that develops prediction models using **Artificial Neural Networks** (**ANNs**). ANNs, simply called **Neural Networks** (**NNs**), are computations that are loosely based on how human brains with neurons work to process information. ANNs consist of neurons, which are types of nodes that are interconnected with other neurons. These neurons transmit information among themselves; this gets processed down the NN to eventually arrive at a result.

DL is one of the most powerful ML techniques and is used to train models that are highly configurable and can support predictions for large and complicated datasets. DL models can be supervised, semi-supervised, or unsupervised, depending on their configuration.

There are various types of ANNs:

- **Recurrent Neural Network (RNN)**: RNN is a type of NN where the connections between the various neurons of the NN can form a directed or undirected graph. This type of network is cyclic since the outputs of the network are fed back to the start of the network and contribute to the next cycle of predictions. The following diagram shows an example of an RNN:

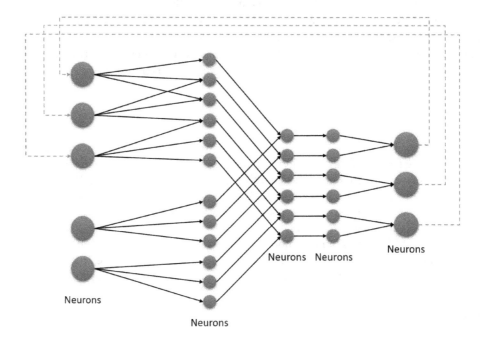

Figure 5.35 – RNN

As you can see, the values from the last nodes in the NN are fed to the starting nodes of the network as inputs.

- **Feedforward NN**: A feedforward neural network is similar to an RNN, with the only difference being that the network of nodes does not form a cycle. The following diagram shows an example of a feedforward neural network:

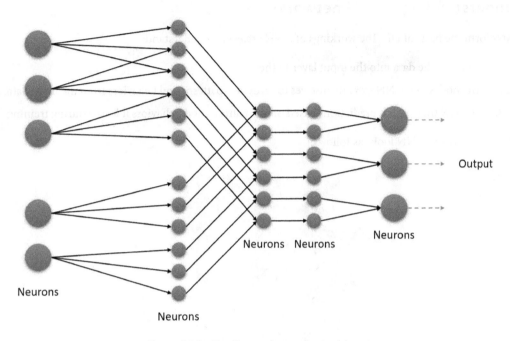

Figure 5.36 – Feedforward neural network

As you can see, this type of NN is unidirectional. This is the simplest type of ANN. A feedforward neural network is also called **Deep Neural Network (DNN)**.

H2O's DL is based on a multi-layer free forward ANN. It is trained on **stochastic gradient descent** using **backpropagation**. There are plenty of different types of DNNs that H2O can train. They are as follows:

- **Multi-Layer Perceptron (MLP)**: These types of DNNs are best suited for tabular data.
- **Convolutional Neural Networks (CNNs)**: These types of DNNs are best suited for image data.
- **Recurrent Neural Networks (RNNs)**: These types of DNNs are best suited for sequential data such as voice data or time series data.

It is recommended to use the default DNNs that H2O provides out of the box as configuring a DNN can be very difficult for non-experts. H2O has already preconfigured its implementation of DL to use the best type of DNNs for the given cases.

With these basics in mind, let's dive deeper into understanding how DL works.

Understanding neural networks

NNs form the basis of DL. The workings of a NN are easy to understand:

1. You feed the data into the input layer of the NN.

2. The nodes in the NN train themselves to recognize patterns and behaviors from the input data.

3. The NN then makes predictions based on the patterns and behaviors it learns during training.

The structure of an NN looks as follows:

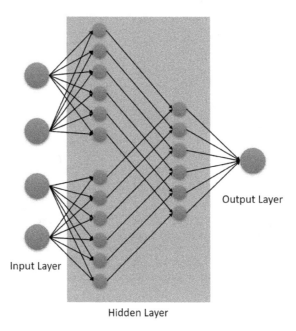

Figure 5.37 – Structure of an NN

There are three essential components of an NN:

- **The input layer**: The input layer consists of multiple sets of neurons. These neurons are connected to the next layer of neurons, which reside in the hidden layer.

- **The hidden layer**: Within the hidden layer, there can be multiple layers of neurons, all of which are interconnected layer by layer.

- **The output layer**: The output layer is the final layer in the NN that makes the final calculations to compute the final prediction values in terms of probability.

The learning process of the NN can be broken down into two components:

- **Forward propagation**: As the name suggests, forward propagation is where the information flows from the input layer to the output layer through the middle layer. The following diagram shows forward propagation in action:

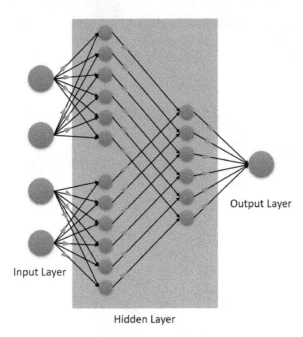

Figure 5.38 – Forward propagation in an NN

The neurons within the middle layer are connected via **channels**. These channels are assigned numerical values called **weights**. Weights determine how important the neuron is in terms of its value contributing to the overall prediction. The higher the value of the weight, the more important that node is when making predictions.

The input values from the input layer are multiplied by these weights as they pass through the channels and their sum is sent as inputs to the neurons in the hidden layer. Each neuron in the hidden layer is associated with a numerical value called a **bias**, which is added to the input sum.

This weighted value is then passed to a non-linear function called the activation function. The activation function is a function that decides if the particular neuron can pass its calculated weight value onto the next layer of the neuron or not, depending on the equation of the non-linear function. The bias is a scalar value that shifts the activation function either to the left or right of the graph for corrections.

This flow of information continues to the next layer of neurons in the hidden layer, following the same process of multiplying the weight of the channels and passing the input to the next activation function of the node.

Finally, in the output layer, the neuron with the highest value determines what the prediction value is, which is a form of probability.

- **Backpropagation**: Backpropagation works the same way as forward propagation except that it works in the reverse direction. Information is passed from the output layer to the input layer through the hidden layer in a reverse manner. The following diagram will give you a better understanding of this:

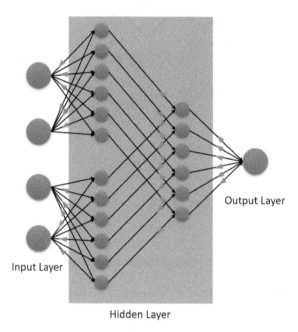

Figure 5.39 – Backpropagation in NN

It may be counterintuitive to understand how backpropagation can work since it works in a reverse manner from the output to the input, but it is a concept that makes DL so powerful for ML. It is through backpropagation that NNs can learn by themselves.

The way it does this is pretty simple. In backpropagation learning, the NN will calculate the magnitude of error between the expected value and the predicted value and evaluate its performance by mapping it onto a **loss function**. The loss function is a function that calculates the deviance between the predicted and expected values. This deviation value is the information that helps the NN adjust its biases and weights in the hidden layer to improve its performance and make better predictions.

Congratulations – we have just gotten a basic glimpse of how DL trains ML models!

> **Tip**
>
> DL is one of the most sophisticated fields in ML, as well as AI as a whole. It's a vast area of ML specialization where data scientists spend a lot of time researching and understanding the problem statement and the data they are working with so that they can correctly tune their DL NN. The mathematics behind it is also very complex and as such deserves a dedicated book. So, if you are interested in mathematics and want to excel in the art of DL, feel free to explore the ML algorithm in depth as every step in understanding it will make you that much more of an expert ML engineer. You can find more information about how H2O performs DL training at `https://docs.h2o.ai/h2o/latest-stable/h2o-docs/data-science/deep-learning.html`.

Summary

In this chapter, we understood the different types of prediction problems and how various algorithms aim to solve them. Then, we understood how the different ML algorithms are categorized into supervised, unsupervised, semi-supervised, and reinforcement based on their method of learning from data. Once we had an understanding of the overall problem domain of ML, we understood that H2O AutoML trains only supervised learning ML algorithms and can solve prediction problems in this domain specifically.

Then, we understood which algorithms H2O AutoML trains starting with GLM. To understand GLM, we understood what linear regression is and how it works and what assumptions about the normal distribution of data it has to make to be effective. With these basics in mind, we understood how GLM is generalized to be effective, even if these assumptions of linear regression are met, which is a common case in real life.

Then, we learned about DRF. To understand DRF, we understood what decision trees are – that is, the basic building blocks of DRF. Then, we learned that multiple decision trees with their ensembled learnings are better ML models than a normal decision tree – that is how Random Forest works. Building on top of this, we learned how DRF adds more randomization in the form of XRT to make the algorithm all the more effective with low variance and bias.

After that, we learned about GBM. We learned how GBM is similar to DRFs but that it has a slightly different way of learning. We understood how GBM sequentially builds decision trees and slowly minimizes error by learning from its residuals from previous decision tree prediction aggregates.

Finally, we learned what DL is. We understood how NNs are the building blocks of DL and their different types. We also understood how NNs perform backpropagation learning from its results and self-learn and improve the model by adjusting the weights and biases of the neurons in the middle layer.

This chapter gave you a brief conceptual understanding of how the various ML algorithms are trained by H2O AutoML without diving too deep into the mathematics. However, ML enthusiasts who want to become experts in the field of ML and wish to work on complex ML problems are strongly encouraged to understand the math behind the wonderful world of ML algorithms. It is the culmination of years of research and effort by scientists and enthusiasts such as yourselves that we have the capability today to potentially predict the future with the help of machines.

In the next chapter, we shall dive deep into understanding how you can understand if an ML model is performing optimally or not using different statistical measurements and other metrics that explain more about ML model performance.

Understanding H2O AutoML Leaderboard and Other Performance Metrics

When we train ML models, the statistical nuances of different algorithms often make it difficult to compare one model with another model that is trained using a different algorithm. From a professional standpoint, you will eventually need to select the right model to solve your ML problem. So, the question arises: how do you compare two different models solving the same ML problem and decide which one is better?

This is where model performance metrics come in. Model performance metrics are certain numerical metrics that give an accurate measurement of a model's performance. The performance of a model can mean various things and can also be measured in several ways. The way we evaluate a model, whether it is a classification model or a regression model, only differs by the metrics that we use for that evaluation. You can measure how accurately the model classifies objects by measuring the number of correct and incorrect predictions. You can measure how accurately the model predicted a stock price and note the magnitude of the error between the predicted value and the actual value. You can also compare how the model fairs with outliers in data.

H2O provides plenty of model performance measuring techniques. Most of them are automatically calculated and stored as the model metadata whenever a model is trained. H2O AutoML further automates the selection of models as well. It does so by presenting you with a leaderboard comparing the different performance metrics of the trained models. In this chapter, we will explore the different performance metrics that are used in the AutoML leaderboard, as well as some additional metrics that are important for users to know.

We shall explore these performance metrics according to the following sections:

- Exploring the H2O AutoML leaderboard performance metrics
- Exploring other important performance metrics

By the end of this chapter, you should understand how a model's performance is measured and how we can use these metrics to get an understanding of its prediction behavior.

So, let's begin by exploring and understanding the H2O AutoML leaderboard performance metrics.

Exploring the H2O AutoML leaderboard performance metrics

In *Chapter 2, Working with H2O Flow (H2O's Web UI)*, once we trained the models on a dataset using H2O AutoML, the results of the models were stored in a leaderboard. The leaderboard was a table containing the model IDs and certain metric values for the respective models (*see Figure 2.33*).

The leaderboard ranks the models based on a default metric, which is ideally the second column in the table. The ranking metrics depend on what kind of prediction problem the models are trained on. The following list represents the ranking metrics used for the respective ML problems:

- For binary classification problems, the ranking metric is **AUC**.
- For multi-classification problems, the ranking metric is the **mean per-class error**.
- For regression problems, the ranking metric is **deviance**.

Along with the ranking metrics, the leaderboard also provides some additional performance metrics for a better understanding of the model quality.

Let's try to understand these performance metrics, starting with the mean squared error.

Understanding the mean squared error and the root mean squared error

Mean Squared Error (MSE), also called **Mean Squared Deviation (MSD)**, as the name suggests, is a metric that measures the mean of the squares of errors of the predicted value against the actual value.

Consider the following regression scenario:

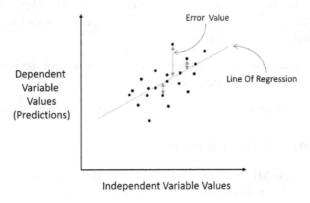

Figure 6.1 – The MSE in a regression scenario

This is a generic regression scenario where the line of regression passes through the data points plotted on the graph. The trained model makes predictions based on this line of regression. The error values show the difference between the actual value and the predicted value, which lies on the line of regression, as denoted by the red lines. These errors are also called residuals. When calculating the MSE, we square these errors to remove any negative signs, as we are only concerned with the magnitude of the error and not its direction. The squaring also gives more weight to larger error values. Once all the squared errors have been calculated for all the data points, we calculate the mean, which gives us the final MSE value.

The MSE is a metric that tells you how close the line of regression is to the data points. Accordingly, the fewer error values the line of regression has against the data points, the lower your MSE value will be. Thus, when comparing the MSE of different models, the model with the lower MSE is ideally the more accurate one.

The mathematical formula for the MSE is as follows:

$$MSE = 1/n * \Sigma(actual\ values - predicted\ value)^2$$

Here, n would be the number of data points in the dataset.

The **Root Mean Squared Error** (**RMSE**), as the name suggests, is the root value of the MSE. So accordingly, its mathematical formula is as follows:

$$RMSE = \sqrt{MSE} = \sqrt{(1/n * \Sigma(actual\ values - predicted\ value)^2)}$$

The difference between the MSE and the RMSE is straightforward. While the MSE is measured in the squared units of the response column, the RMSE is measured in the same units as the response column.

For example, if you have a linear regression problem that predicts the price of its stock in terms of dollars, the MSE measures the errors in terms of squared dollars, while RMSE measures the error value as just dollars. Hence, the RMSE is often used over the MSE, as it is slightly easier to interpret the model quality from the RMSE than the MSE.

Congratulations – you are now aware of what MSE and RMSE metrics are and how they can be used to measure the performance of a regression model.

Let's move on to the next important performance metric, which is the confusion matrix.

Working with the confusion matrix

A classification problem is an ML problem where the ML model tries to classify the data inputs into the pre-specified classes. What makes the performance measurement of classification models different from regression models is that in classification problems, there is no numeric magnitude of the error between the predicted value and the actual value. The predicted value is either correctly classified into the right class or it is incorrectly classified. To measure model performance for classification problems, data scientists rely on certain performance metrics that are derived from a special type of matrix called a **confusion matrix**.

A confusion matrix is a tabular matrix that summarizes the correctness of the prediction results of a classification problem. The matrix presents the count of correct and incorrect predicted values alongside each other, as well as breaking them down by each class. This matrix is called a confusion matrix, as it shows how confused the model is when classifying the values.

Consider the example of the heart disease prediction dataset we used. It is a binary classification problem where we want to predict whether a person with certain health conditions is likely to suffer from heart disease or not. In this case, the prediction is either **Yes**, also called a **positive classification**, meaning the person is likely to suffer from heart disease, or **No**, also called a **negative classification**, meaning the person is not likely to suffer from heart disease.

The confusion matrix of this scenario will be as follows:

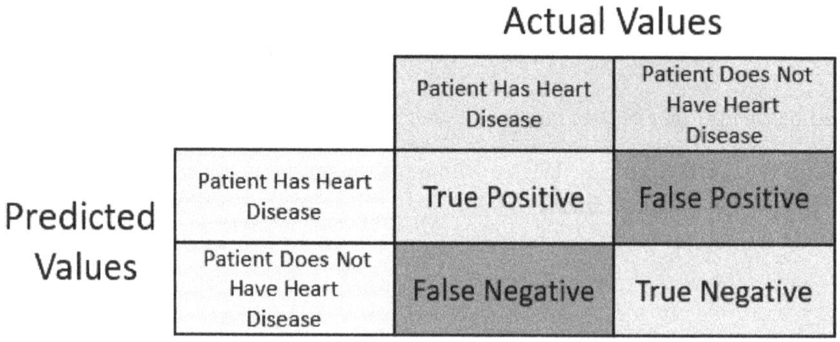

		Actual Values	
		Patient Has Heart Disease	Patient Does Not Have Heart Disease
Predicted Values	Patient Has Heart Disease	True Positive	False Positive
	Patient Does Not Have Heart Disease	False Negative	True Negative

Figure 6.2 – A binomial confusion matrix

The rows of the confusion matrix correspond to the classifications predicted by the model. The columns of the confusion matrix correspond to the actual class values of the model.

In the top-left corner of the matrix, we have **true positives** – these are the number of **Yes** actuals that were correctly predicted as Yes. In the top-right corner, we have the **false positives** – these are the number of Yes actuals that were incorrectly predicted as **No**. In the bottom-left corner, we have **false negatives** – these are the number of No actuals that were incorrectly predicted as Yes values. And finally, we have the **true negative** – these are the number of No actuals that were correctly predicted as No.

The confusion matrix for a multinomial classification with six possible classes will look as follows:

		Actual Values					
		Class A	Class B	Class C	Class D	Class E	Class F
Predicted Values	Class A	Correct Prediction	Incorrect Prediction	Incorrect Prediction	Incorrect Prediction	Incorrect Prediction	Incorrect Prediction
	Class B	Incorrect Prediction	Correct Prediction	Incorrect Prediction	Incorrect Prediction	Incorrect Prediction	Incorrect Prediction
	Class C	Incorrect Prediction	Incorrect Prediction	Correct Prediction	Incorrect Prediction	Incorrect Prediction	Incorrect Prediction
	Class D	Incorrect Prediction	Incorrect Prediction	Incorrect Prediction	Correct Prediction	Incorrect Prediction	Incorrect Prediction
	Class E	Incorrect Prediction	Incorrect Prediction	Incorrect Prediction	Incorrect Prediction	Correct Prediction	Incorrect Prediction
	Class F	Incorrect Prediction	Incorrect Prediction	Incorrect Prediction	Incorrect Prediction	Incorrect Prediction	Correct Prediction

Figure 6.3 – A multinomial confusion matrix

Using the confusion matrix of two classification models, you can compare the number of true positives and true negatives that were predicted by the individual algorithms and select the one with the greater number of correct predictions as the better model.

Despite being very easy to interpret the prediction quality of a model using the confusion matrix, it is still difficult to compare two or more models solely based on the number of true positives and true negatives.

Consider a scenario where you want to classify some medical records to identify whether the patient has a brain tumor. Let's assume that a specific model's confusion matrix has a high number of true positives and true negatives compared to other models and also has a high number of false positives. In this case, the model will incorrectly flag a lot of normal medical records as indicative of a potential brain tumor. This might result in hospitals making incorrect decisions and performing risky surgeries that were never needed. In such a scenario, models with less accuracy but the smallest number of false positives are preferable.

Hence, more sophisticated metrics are developed on top of the confusion matrix. They are as follows:

- **Accuracy**: Accuracy is a metric that measures the number of correctly predicted positive and negative predictions against the total number of predictions made. This is calculated as follows:

$$Accuracy = \frac{TP + TN}{TP + FP + TN + FN}$$

  Here, the abbreviations stand for the following:

  - **TP** stands for True Positive.

  - **TN** stands for True Negative.

  - **FP** stands for False Positive.

  - **FN** stands for False Negative.

  This metric is useful when you want to compare how well a classification model correctly makes predictions, irrespective of whether the prediction value is positive or negative.

- **Precision**: Precision is a metric that measures the number of correct positive predictions made compared to the total number of positive predictions made. This is calculated as follows:

$$Precision = \frac{True\ Positives}{True\ Positives + False\ Positives}$$

  This metric is especially useful when measuring the performance of a classification model that is trained on data with a high number of negative results and only a few positive results. Precision is not affected by the imbalance of positive and negative classification values, as it only considers positive values.

- **Sensitivity or recall**: Sensitivity, also known as recall, is a probability measurement for how well a model can predict true positives. Sensitivity is measured by identifying what percentage of predictions were correctly identified as positive in a binomial classification. This is calculated as follows:

$$Sensitivity = \frac{True\ Positives}{True\ Positives + False\ Negatives}$$

  If your classification ML problem aims to accurately identify all the positive predictions, then the sensitivity of the model should be high.

- **Specificity**: While sensitivity is the probability measurement of how well a model can predict true positives, specificity is measured by identifying what percentage of predictions were correctly identified as negative in a binomial classification. This is calculated as follows:

$$Specificity = \frac{True\ Negatives}{True\ Negatives + False\ Positives}$$

If your classification ML problem aims to accurately identify all the negative predictions, then the specificity of the model should be high.

There is always a trade-off between sensitivity and specificity. A model with high sensitivity will often have very low specificity and vice versa. Thus, the context of the ML problem plays a very important part in deciding whether you want a model with high sensitivity or high specificity to solve the problem.

For multinomial classification, you calculate the sensitivity and specificity for each class type. For sensitivity, your true positives will remain the same, but the false negatives will change depending on the number of incorrect predictions made for that class. Similarly, for specificity, the true negatives will remain the same – however, the false positives will change depending on the number of incorrect predictions made for that class.

Now that you have understood how a confusion matrix is used for measuring classification models and how sensitivity and specificity are built on top of it, let's now move on to the next metric, which is the receiver operating characteristic curve and its area under the curve.

Calculating the receiver operating characteristic and its area under the curve (ROC-AUC)

Another good way of comparing classification models is via a visual representation of their performance. One of the most widely used visual evaluation metrics is the **Receiver Operating Characteristic** and its **Area Under the Curve (ROC-AUC)**.

The ROC-AUC metric is split into two concepts:

- **The ROC curve**: This is the graphical curve plotted on a graph that summarizes the model's classification ability at various thresholds. The threshold is a classification value that separates the data points into different classes.

- **AUC**: This is the area under the ROC curve that helps us compare which classification algorithm performed better depending on whose ROC curve covers the most area.

Let's consider an example to better understand how ROC-AUC can help us compare classification models. Refer to the following sample dataset:

Weight (kgs)	Obese
70	0
75	0
80	0
85	0
90	1
95	0
100	1
105	0
110	1
115	1
120	1

Figure 6.4 – An obesity dataset

This dataset has two columns:

- **Weight (kgs)**: This is a numerical column that contains the weight of a person in kilograms

- **Obese**: This is a categorical column that contains either **1** or **0**, where **1** indicates the person is obese and **0** indicates that the person is not obese

Let's plot this dataset onto a graph where **Weight**, being the independent variable, is on the x-axis, and **Obese**, being the dependent variable, is on the y-axis. This simple dataset on a graph will look as follows:

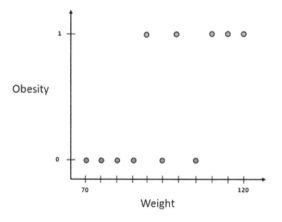

Figure 6.5 – The plotted obesity dataset

Let's fit a classification line through this data using one of the simplest classification algorithms called **logistic regression**. Logistic regression is an algorithm that predicts the probability that a given sample of data belongs to a certain class. In our example, the algorithm will predict the probability of whether the person is obese or not depending on their weight.

The logistic regression line will look as follows:

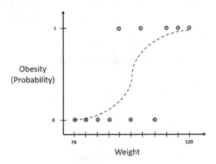

Figure 6.6 – A plotted obesity dataset with a classification line

Note that since logistic regression predicts the probability that data might belong to a certain class, we have converted the y-axis into the probability that the person is obese.

During prediction, we will first plot the sample weight data of the person on the x-axis. We will then find its respective y value on the line of classification. This value is the probability that the respective person is obese.

Now, to classify whether the person is obese or not, we will need to decide what the probability **cut-off line** that separates obese and not obese will be. This cut-off line is called the **threshold**. Any probability value above the threshold can be categorized as obese and any value below it can be categorized as not obese. The threshold can be any value between 0 and 1:

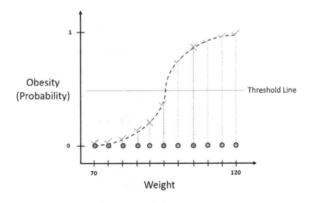

Figure 6.7 – An obesity dataset classification with a threshold line

As you can see from the diagram, multiple values are incorrectly classified. This is bound to happen in any classification problem. So, to keep a track of the correct and incorrect classification, we will create a confusion matrix and calculate sensitivity and specificity to evaluate how well the model performs for the selected threshold.

But as mentioned previously, there can be many thresholds for classification. Thresholds with high values will minimize the number of false positives but the trade-off is that classification for that class will become stricter, leading to more false negatives. Similarly, if the threshold value is too low, then we will end up with more False Positives.

Which threshold performs best depends on your ML problem. However, a comparative study of different thresholds is needed to find a suitable value. Since you can create any number of thresholds, you will end up creating plenty of confusion matrices eventually. This is where the ROC-AUC metric comes in.

The ROC-AUC metric summarizes the performance of the model at different thresholds and plots them on a graph. In this graph, the *x*-axis is the False Positive rate, which is **1 - specificity**, while the *y*-axis is the true positive rate, which is nothing but **sensitivity**.

Let's plot the ROC graph for our sample dataset. We will start by using a threshold that classifies all samples as obese. The threshold on the graph will look as follows:

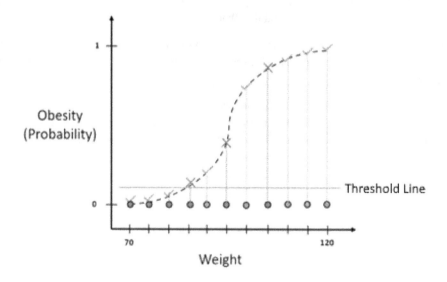

Figure 6.8 – A plotted obesity classification with a very low threshold

We will now need to calculate the sensitivity (and 1 - specificity) values needed to plot our ROC curve, so accordingly, we will need to create a confusion matrix first. The confusion matrix for this threshold will look as follows:

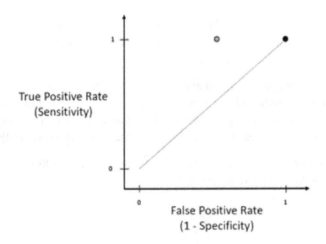

$$\text{Sensitivity} = \frac{TP}{TP + FN} = \frac{5}{5 + 0} = 1$$

$$1 - \text{Specificity} = 1 - \frac{TN}{TN + FP} = 1 - \frac{3}{3 + 3} = 0.5$$

Figure 6.9 – A confusion matrix with sensitivity and 1 - specificity

Calculating the sensitivity and 1 - specificity values using the formula mentioned previously, we get a sensitivity equal to **1** and a 1 - specificity equal to **0.5**. Let's plot this value in the ROC graph. The ROC graph will look as follows:

Figure 6.10 – The ROC graph

The blue line in the diagram indicates that the sensitivity is equal to the 1 - specificity – in other words, the true positive rate is equal to the False Positive rate. Any ROC points on this line indicate that the model trained using this threshold has an equal likelihood of predicting a correct positive as predicting an incorrect positive. So, to find the best threshold, we aim to find a ROC point that has as high a sensitivity as possible and as low a 1 - specificity as possible. This would indicate that the model has a high likelihood of predicting a correct positive prediction and a much smaller likelihood of predicting an incorrect positive prediction.

Let's now raise the threshold and repeat the same process to calculate the ROC value for this new threshold. Let's assume this new threshold has a sensitivity of 1 and a 1 - specificity of 0.25. Plotting this value in the ROC graph, we get the following result:

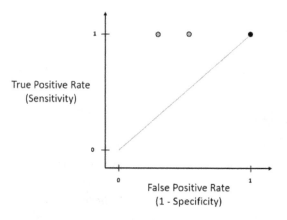

Figure 6.11 – A ROC graph with the new threshold

The new ROC value for the new threshold is on the left side of the blue line and also of the previous ROC point. This indicates that it has a lower false positive rate compared to the previous threshold. Thus, the new threshold is better than the previous one.

Raising the threshold value way too high will make the model predict that all the values are not obese. Basically, it will incorrectly predict all the values as false, increasing the number of false negatives. Based on the sensitivity equation, the higher the number of false negatives, the lower the sensitivity. So, this will eventually lower your sensitivity, reducing the model's ability to predict the true positives.

We repeat this same process for different threshold values and plot their ROC values on the ROC graph. If we connect all these dots, we get the ROC curve:

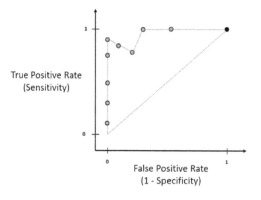

Figure 6.12 – The ROC graph with a ROC curve

Just by looking at the ROC graph, you can identify which threshold values are better than the others, and depending on how many false positive predictions your ML problem can tolerate, you can select the ROC point with the right false positive rate as your final threshold value reference. This explains what the ROC curve does.

Now, suppose you have another algorithm trained with different thresholds and you plot its ROC points on this same graph. Assume the plots look as follows:

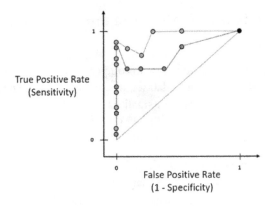

Figure 6.13 – A ROC graph with multiple ROC curves

How would you compare which algorithm performed better? Which threshold is the optimum one for that algorithm's model?

This is where AUC helps us. AUC is nothing but the area under the ROC curve. The whole ROC graph will have a total area of *1*. The red line splits the area into half, so ideally, all potentially good algorithms should have an AUC greater than 0.5. The greater the AUC is, the better the algorithm is:

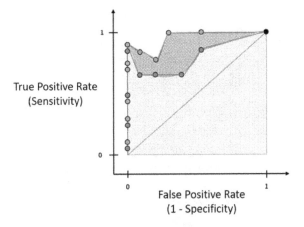

Figure 6.14 – The AUC of the ROC curve

Just by visualizing this, you can see which algorithm is better from its AUC. Similarly, the AUC values help engineers and scientists identify which algorithm to choose and which threshold to use as the optimal ML model for classification.

Congratulations, you just understood how the ROC-AUC metric works and how it can help you compare model performance. Let's now move on to another similar performance metric called the **Precision-Recall curve (PR curve)**.

Calculating the precision-recall curve and its area under the curve (AUC-PR)

With ROC-AUC, despite being a very good metric to compare models, there is a minor drawback to relying on it exclusively. In a very imbalanced dataset, where there is a large number of true negative values, the x-axis of the ROC graph will be very small, as specificity has a true negative value as its denominator. This forces the ROC curve toward the left side of the graph, raising the ROC-AUC value toward 1, which is technically incorrect.

This is where the **PR curve** proves beneficial. The PR curve is similar to the ROC curve, the only difference being that the PR curve is a function that uses precision on the y-axis and recall on the x-axis. Neither precision nor recall uses true negatives in their calculation. Hence, the PR curve and its AUC metric are suitable when there is an imbalance in the classes of the dataset that impacts the true negatives during prediction, or when your ML problem is not concerned with true negatives at all.

Let's understand the PR curve further using an example. We will use the same sample obesity dataset that we used for understanding the ROC-AUC curve. The process of plotting the records of the dataset on the graph and creating its confusion matrix is the same as for the ROC-AUC curve.

Now, instead of calculating the sensitivity and 1 – specificity from the confusion matrix, this time, we shall calculate the precision and recall values:

		Actual Values	
		Patient is Obese	Patient is not Obese
Predicted Values	Patient is Obese	5	3
	Patient is not Obese	0	3

$$Precision = \frac{TP}{TP + FP} = \frac{5}{5 + 3} = 0.625$$

$$Recall = \frac{TP}{TP + FN} = \frac{5}{5 + 0} = 1$$

Figure 6.15 – Calculating the precision and recall values

As you can see from the preceding diagram, we got precision values of **0.625** and recall values of **1**. Let's plot these values onto the PR graph as shown in the following diagram:

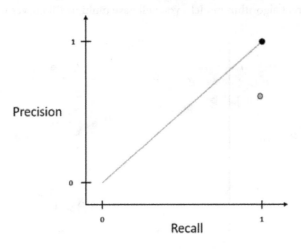

Figure 6.16 – The PR graph

Similarly, by moving the threshold line and creating the new confusion matrix, the precision and recall values will change based on the distribution of the predictions in the confusion matrix. We repeat this same process for different threshold values, calculate the precision and recall values, and then plot them onto the PR graph:

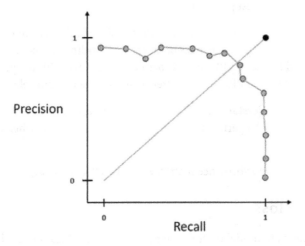

Figure 6.17 – A PR graph and its PR curve

The blue line that joins all the points is the PR curve. The point that represents a threshold value closest to the black point, so closest to having a precision value of 1 and a recall value of 1, is the ideal classifier.

When comparing different algorithm models, you will have multiple PR curves in the PR graph:

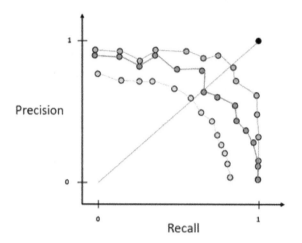

Figure 6.18 – A PR graph with multiple PR curves

The preceding diagram shows you the multiple PR curves that can be plotted on the same graph to give you a better comparative view of the performances of different algorithms. With one glance, you can see that the algorithm represented by the blue line has a threshold value that is closest to the black point and should ideally be the best performing model.

Just as with ROC-AUC, you can also use the AUC-PR to calculate the area under the PR curves to get a better understanding of the performances of different algorithms. Based on this, you know that the algorithm represented by the red PR curve is better than the one with the yellow curve and the algorithm represented by the blue PR curve is better than both the red and yellow curves.

Congratulations! You now understand the AUC-PR metric in the H2O AutoML leaderboard and how it can be another good model performance metric that you can refer to when comparing models trained by H2O AutoML.

Let's now move on to the next performance metric, which is called log loss.

Working with log loss

Log loss is another important model performance metric for classification models. It is primarily used to measure the performance of binary classification models.

Log loss is a way of measuring the performance of a classification model that outputs classification results in the form of probability values. The probability values can range from 0, which means that the data has zero probability that it belongs to a certain positive class, to 1, which means the data has a 100% chance of belonging to a certain positive class. The log loss value can range from 0 to infinity and the goal of all ML models is to minimize the log loss as much as possible. Any model with a log loss value as close to 0 as possible is regarded as the better performing model.

Log loss calculation is entirely statistical. However, it is important to understand the intuition behind the mathematics to better understand its application when comparing model performances.

Log loss is a metric that measures the divergence of the predicted probability from the actual value. So, if the predicted probability diverges very little from the actual value, then your log loss value will be forgiving – however, if the divergence is greater, the log loss value will be that much more punishing.

Let's start by understanding what prediction probability is. We shall use the same obesity dataset that we used for the ROC-AUC curve. Assume we ran a classification algorithm that calculated the prediction probability that the person is obese and let's add those values to a column in the dataset, as seen in the following screenshot:

Weight (kgs)	Obese	Prediction Probability
70	0	0.29
75	0	0.35
80	0	0.40
85	0	0.52
90	1	0.54
95	0	0.60
100	1	0.62
105	0	0.69
110	1	0.75
115	1	0.007
120	1	0.0002

Figure 6.19 – The obesity dataset with the prediction probabilities added

We will have a certain threshold value that decides what the prediction probability value has to be for us to classify the data as obese or not obese. Let's assume the threshold is 0.5 – in this case, a prediction probability value above 0.5 is classified as obese and anything below it is classified as not obese.

We now calculate the log loss value of each data point. The equation for calculating the log loss of each record is as follows:

$$Logloss\ of\ record = -[y\log(p) + (1 - y)\log(1 - p)]$$

Here, the equation can be broken down as follows:

- **y** is the actual classification value, that is, *0* or *1*.

- **p** is the prediction probability.

- **log** is the natural logarithm of the number.

In our example, since we are using the obese class as a reference, we shall set *y* to *1*. Using this equation, we calculate the log loss value of individual data values as follows:

Weight (kgs)	Obese	Prediction Probability		Log loss
70	0	0.29		0.14
75	0	0.35		0.18
80	0	0.40		0.22
85	0	0.52		0.31
90	1	0.54		0.26
95	0	0.60		0.39
100	1	0.62		0.20
105	0	0.69		0.50
110	1	0.75		0.12
115	1	0.007		2.15
120	1	0.0002		3.69

Figure 6.20 – An obesity dataset with the log loss values per record

Now, let's plot these values into a log loss graph where we set the log loss values on the *y*-axis and the prediction probability on the *x*-axis:

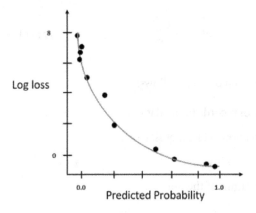

Figure 6.21 – A log loss graph where y = 1

You will notice that the log loss value exponentially rises as the predicted probability diverges away from the actual value. The lesser the divergence, the less the increase in log loss will be. This is what makes log loss a good comparison metric, as it compares not only which model is good or bad but also how good or bad it is.

Similarly, if you wanted to use the not obese class as a reference for log loss, then you would inverse the prediction probabilities, calculate the log loss values, and plot the graph, or you could just calculate the log loss value by setting *y* to 0 and use the log loss values calculated to plot the log loss graph. This graph will be a mirror image of the previous graph (*see Figure 6.21*):

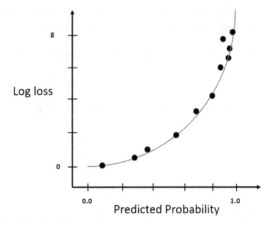

Figure 6.22 – A log loss graph where y = 0

The log loss value of the model, also called the **skill** of the model, is the average of the log loss values for all the records in the dataset. Accordingly, the equation for the log loss of the model is as follows:

$$Logloss\ of\ model = -\frac{1}{n}\sum_{i=1}^{n}[y\log(p) + (1-y)\log(1-p)]$$

Here, the equation can be broken down as follows:

- **n** is the total number of records in the dataset.
- **y** is the actual classification value, that is, *0* or *1*.
- **p** is the prediction probability.
- **log** is the natural logarithm of the number.

In an ideal world, a model with perfect scoring capabilities and skills is said to have a log loss equal to *0*. To correctly apply log loss to compare models, both models must be trained using the same dataset.

Congratulations! We have just covered how log loss is statistically calculated. In the next section, we shall explore some other important metrics that are not a part of the H2O AutoML leaderboard but are nonetheless important in terms of understanding model performance.

Exploring other model performance metrics

The H2O AutoML leaderboard summarizes the model performances based on certain commonly used important metrics. However, there are still plenty of performance metrics in the field of ML that describe different skills of the ML model. These skills can often be the deciding factor in what works best for your given ML problem and hence, it is important that we are aware of how we can use these different metrics. H2O also provides us with these metrics values by computing them once training is finished and storing them as the model's metadata. You can easily access them using built-in functions.

In the following subsections, we shall explore some of the other important model performance metrics, starting with F1.

Understanding the F1 score performance metric

Precision and recall, despite being very good metrics to measure a classification model's performance, have a trade-off. Precision and recall cannot both have high values at the same time. If you increase the precision by adjusting your classification threshold, then it impacts your recall, as the number of false negatives might increase, reducing your recall value, and vice versa.

The precision metric works to minimize incorrect predictions, while the recall metric works to find the greatest number of positive predictions. So, technically, we need to find the right balance between these two metrics.

This is where the **F1 score** performance metric comes into the picture. The F1 score is a metric that tries to maximize both precision and recall at the same time and gives an overall score for the model's performance.

The F1 score is the harmonic mean of the precision and recall values. **The harmonic mean** is just one of the variations for calculating the mean of values. With the harmonic mean, we calculate the reciprocal of the arithmetic mean of the reciprocals of all the observations. The reason we use a harmonic mean for calculating the F1 score is that using a general arithmetic mean would lead to the equation giving equal importance to all degrees of error. The harmonic mean, on the other hand, punishes high values of error by lowering the F1 score accordingly. This is the reason why the harmonic mean is used to generate the F1 score, as the score value calculated gives a better representation of the model's performance.

The F1 score ranges from 0 to 1, where 1 indicates that the model has perfect precision and recall values, while 0 indicates that either the precision or recall value is 0.

The equation for calculating the F1 score is as follows:

$$F1 = \frac{2 * Precision * Recall}{Precision + Recall}$$

Let's take the example of a confusion matrix as follows:

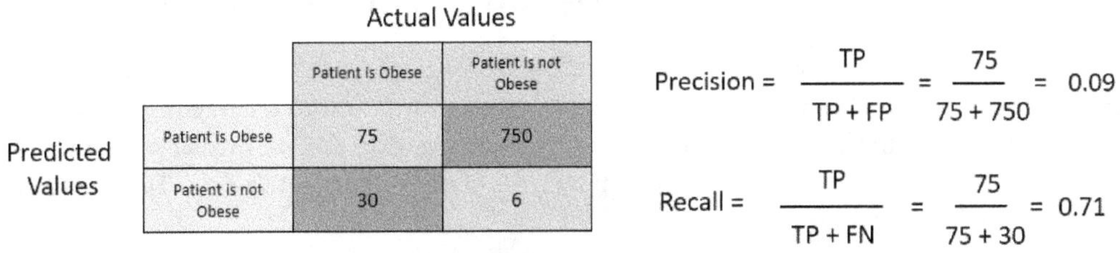

Figure 6.23 – An example confusion matrix with precision and recall values

Let's calculate the precision value for the matrix:

$$Precision = \frac{True\ Positives}{True\ Positives + False\ Positives} = \frac{75}{75 + 750} = 0.09$$

Similarly, lets now calculate the recall value of the matrix:

$$Recall = \frac{True\ Positives}{True\ Positives + False\ Negatives} = \frac{75}{75 + 30} = 0.71$$

Now, plugging the precision and recall values into the F1 score equation, we get the following:

$$F1 = \frac{2 * Precision * Recall}{Precision + Recall} = \frac{2 * 0.09 * 0.71}{0.09 + 0.71} = 0.15$$

We got an F1 score of *0.15*. You can now compare the performance of another model by similarly calculating its F1 score and comparing it to this one. If the F1 score of the new model is greater than *0.15*, then that model is more performant than this one.

The benefit of the F1 score is that when comparing classification models, you don't need to balance the precision and recall values of multiple models and make a decision based on the comparisons between two contrasting metrics. The F1 score summarizes the optimum values of precision and recall, making it easier to identify which model is better.

Despite being a good metric, the F1 score still has certain drawbacks. Firstly, the F1 score does not consider true negatives when calculating the score. Secondly, the F1 Score does not adequately capture the performance of a multi-class classification problem. You can technically calculate the F1 score for multi-class classification problems using macro-averaging – however, there are better metrics that can be used instead.

Let's look at one such metric that overcomes the drawbacks of the F1 score, which is the absolute Matthews correlation coefficient.

Calculating the absolute Matthews correlation coefficient

Consider an example where we are trying to predict, based on a given fruit's size, whether it is a grape or a watermelon. We have 200 samples, out of which 180 are grapes and 20 are watermelon. Pretty simple, yes – the bigger the size, the more likely it is to be a watermelon, while a smaller size indicates it is a grape. Assume that we trained a classifier taking the grape as the positive class. This classifier was able to classify the fruits as follows:

		Actual Values	
		Fruit is grape	Fruit is not a grape
Predicted Values	Fruit is grape	163	17
	Fruit is not a grape	12	8

Figure 6.24 – A fruit classification confusion matrix with the grape as the positive class

Let's quickly calculate the scalar classification metrics that we previously covered.

The accuracy of the classifier will be as follows:

$$Accuracy = \frac{TP + TN}{TP + FP + TN + FN} = \frac{163 + 8}{163 + 17 + 8 + 12} = 0.85$$

The precision of the classifier will be as follows:

$$Precision = \frac{True\ Positives}{True\ Positives + False\ Positives} = \frac{163}{163 + 17} = 0.90$$

The recall of the classifier will be as follows:

$$Recall = \frac{True\ Positives}{True\ Positives + False\ Negatives} = \frac{163}{163 + 12} = 0.93$$

The F1 score of the classifier will be as follows:

$$F1 = \frac{2 * Precision * Recall}{Precision + Recall} = \frac{2 * 0.9 * 0.93}{0.9 + 0.93} = 0.91$$

Based on these metric values, it seems that our classifier is performing really well when making a prediction for grapes. So, what if we want to predict watermelons instead? Let's change the positive class to watermelons instead of grapes. The confusion matrix for this scenario will be as follows:

		Actual Values	
		Fruit is watermelon	Fruit is not a watermelon
Predicted Values	Fruit is watermelon	8	12
	Fruit is not a watermelon	17	163

Figure 6.25 – A fruit classification confusion matrix with watermelon as the positive class

We'll quickly calculate the scalar classification metrics that we previously covered.

The accuracy of the classifier will be as follows:

$$Accuracy = \frac{TP + TN}{TP + FP + TN + FN} = \frac{8 + 163}{8 + 12 + 163 + 17} = 0.85$$

The precision of the classifier will be as follows:

$$Precision = \frac{True\ Positives}{True\ Positives + False\ Positives} = \frac{8}{8 + 12} = 0.4$$

The recall of the classifier will be as follows:

$$Recall = \frac{True\ Positives}{True\ Positives + False\ Negatives} = \frac{8}{8 + 17} = 0.32$$

The F1 score of the classifier will be as follows:

$$F1 = \frac{2 * Precision * Recall}{Precision + Recall} = \frac{2 * 0.4 * 0.32}{0.4 + 0.32} = 0.35$$

As we can see from the metric values, accuracy has remained the same but precision, recall, and the F1 score have drastically gone down. Accuracy, precision, recall, and the F1 score, despite being very good metrics for measuring classification performance, have some drawbacks when there is a class imbalance in the dataset. In our grape and watermelon dataset, we only had 20 samples of watermelon in the dataset but 180 samples of grape. This imbalance in data can cause asymmetry in the metric calculation, which can be misleading.

Ideally, as data scientists and engineers, it is often advisable to keep the data as symmetrical as possible to keep the measurements of these metrics as relevant as possible. However, in a real-world dataset with millions of records, it will be difficult to maintain this symmetry. So, it would be beneficial to have some sort of metric that treats both the positive and negative class as equal and gives an overall picture of the classification model's performance.

This is where the **absolute Matthews Correlation Coefficient (MCC)**, also called the **phi coefficient**, comes into play. The equation for the MCC is as follows:

$$MCC\ (phi\ coefficient) = \frac{TP.TN - FP.FN}{\sqrt{(TP + FP).(TP + FN).(TN + FP).(TN + FN)}}$$

During computation, it treats the actual class and the predicted class as two different variables and identifies the correlation coefficient between them. The correlation coefficient is nothing but a numerical value that represents some statistical relationship between the variables. The higher this correlation coefficient value, the better your classification model is.

The MCC values range from -1 to 1. 1 indicates that the classifier is perfect and will always classify the records correctly. A MCC of 0 indicates that there is no correlation between the classes and the prediction from the model is completely random. -1 indicates that the classifier will always incorrectly classify the records.

A classifier with a MCC value of -1 does not mean the model is bad in any sense. It only indicates that the correlation coefficient between the predicted and the actual class is negative. So, if you just reverse the predictions of the classifier, you will always get the correct classification prediction. Also, the MCC is perfectly symmetrical – thus, it treats all the classes equally to provide a metric that considers the overall performance of the model. Switching the positive and negative classes does not affect the MCC value. Thus, if you just take the absolute value of the MCC, it still does not lose the value's relevance. H2O often uses the absolute value of MCC for an easier understanding of the model's performance.

Let's calculate the MCC value of the fruit classification confusion matrix with grape as the positive class:

$$MCC = \frac{163.8 - 17.12}{\sqrt{(163 + 17).(163 + 12).(8 + 17).(8 + 12)}} = \frac{1100}{3968.62} = 0.277$$

Similarly, let's calculate the MCC value of the fruit classification confusion matrix with watermelon as the positive class:

$$MCC = \frac{8.163 - 12.17}{\sqrt{(8 + 12).(8 + 17).(163 + 12).(163 + 17)}} = \frac{1100}{3968.62} = 0.277$$

As you can see, the MCC value remains the same, at *0.277*, even if we switch the positive and negative classes. Also, A MCC of *0.277* indicates that the predicted class and the actual class are weakly correlated, which is correct considering the classifier was bad at classifying watermelons.

Congratulations, you now understand another important metric called the absolute MCC.

Let's now move to the next performance metric, which is R2.

Measuring the R2 performance metric

R2, also called the coefficient of determination, is a regression model performance metric that aims to explain the relationship between the dependent variable and the independent variable in terms of how much of a change in the independent variable affects the dependent variable.

The value of R2 ranges from 0 to 1, where 0 indicates that the regression line is not correctly capturing the trend in the data and 1 indicates that the regression line perfectly captures the trend in the data.

Let's better understand this metric using a graphical example of a dataset. Refer to the below image for the height-to-weight regression graph:

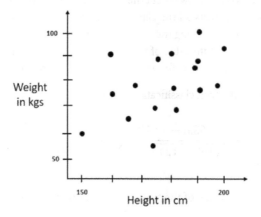

Figure 6.26 – A height-to-weight regression graph

The dataset has two columns:

- **Height**: This is a numerical column that contains the height of a person in centimeters
- **Weight**: This is a numerical column that contains the weight of a person in kilograms

Using this dataset, we are trying to predict someone's weight based on their height.

So firstly, let's use the average of all the weights as a general regression line to predict the weights. Technically, it does make sense, as the majority of people will have an average range of weight for a grown adult body – even though there might be some errors, it is still a plausible way of predicting a person's weight.

If we plot this dataset on a graph, the mean value used for prediction will look as follows:

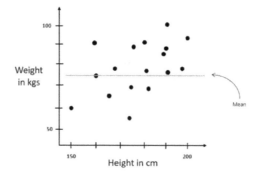

Figure 6.27 – The height-to-weight regression graph with a mean line

As you can see, there is definitely some error between the predicted values of the weight, which is the mean, and the actual values. As mentioned previously, this kind of error is called a residual. Calculating the square of the residuals gives us the squared error. The sum of these squared errors of all the records gives us the variation around the mean line.

Now let's perform linear regression and fit a line through the data so that we get another regression line. This regression line should ideally be a better predictor than using the mean value alone. The regression line on the graph should look as follows:

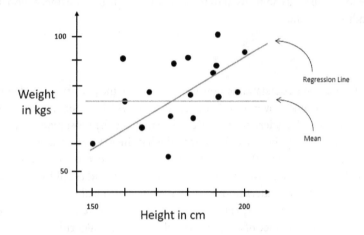

Figure 6.28 – The height-to-weight regression dataset with a regression line

Let's calculate the residual square of errors for this line too – this gives us the variation around the regression line.

Now, we need to figure out a way to identify which line is better, regression or mean, and by how much. This is where R2 can be used to compare the two regression lines. The equation for calculating R2 is as follows:

$$R^2 = 1 - \frac{Sum\ of\ Squares\ of\ Residuals\ of\ the\ Regression\ Line}{Sum\ of\ Squares\ of\ Residuals\ of\ the\ Mean}$$

Let's assume the sum of squares of residuals around the regression line is 7 and the sum of squares of residuals around the mean line is 56. Plugging these values into the R2 equation, we get the following value:

$$R^2 = 1 - \frac{7}{56} = 0.875$$

The value *0.875* is a percentage. This value explains that 87.5 percent of the total variation in the values of *y* is described by the variations in values of *x*. The remaining 12.5 percent may be because of some other factors in the dataset such as muscle mass, fat content, or any other factor.

From an ML perspective, a higher value of R2 indicates that the relationship between the two variables explains the variations in the data and as such, the linear model has captured the pattern of the dataset accurately. A lower R2 value indicates that the linear model has not fully captured the pattern of the dataset and there must be some other factors that contribute to the dataset's pattern.

This sums up how the R2 metric can be used to measure to what degree the linear model is correctly capturing the trends in the data.

Summary

In this chapter, we focused on understanding how we can measure the performance of our ML models and how we can choose one model over the other depending on which is more performant. We started by exploring the H2O AutoML leaderboard metrics since they are the most readily available metrics that AutoML provides out of the box. We first covered what the MSE and the RMSE are, what the difference between them is, and how they are calculated. We then covered what a confusion matrix is and how we calculate accuracy, sensitivity, specificity, precision, and recall from the values in the confusion matrix. With our new understanding of sensitivity and specificity, we understood what a ROC curve and its AUC are, and how they can be used to visually measure the performance of different algorithms, as well as the performance of different models of the same algorithms trained on different thresholds. Building on the ROC-AUC metric, we explored the PR curve, its AUC, and how it overcomes the drawbacks faced by the ROC-AUC metric. And finally, within the leaderboard, we understood what log loss is and how we can use it to measure the performance of binary classification models.

We then explored some important metrics outside of the realm of the leaderboard, starting with the F1 score. We understood how the F1 score incorporates both recall and precision into a single metric. We then understood the MCC and how it overcomes the drawbacks of precision, recall, and the F1 score when measured against imbalanced datasets. And finally, we explored the R2 metrics, which explain the relationship between the dependent variable and the independent variable in terms of how much of a change in the independent variable affects the dependent variable.

With this information in mind, we are now capable of correctly measuring and comparing models to find the most performant model to solve our ML problems. In the next chapter, we shall explore more about the various model explainability features that H2O provides, which give advanced details about a model and its features.

7

Working with Model Explainability

The justification of model selection and performance is just as important as model training. You can have *N* trained models using different algorithms, and all of them will be able to make good enough predictions for real-world problems. So, how do you select one of them to be used in your production services, and how do you justify to your stakeholders that your chosen model is better than the others, even though all the other models were also able to make accurate predictions to some degree? One answer is performance metrics, but as we saw in the previous chapter, there are plenty of performance metrics and all of them measure different types of performance. Choosing the correct performance metric boils down to the context of your ML problem. What else can we use that will help us choose the right model and also further help us in justifying this selection?

The answer to that is visual graphs. Human beings are visual creatures and, as such, a picture speaks a thousand words. A good graph can explain more about a model than any metric number. The versatility of graphs can be very useful in explaining the model's behavior and how it fits as a solution to our ML problem.

H2O's explainability interface is a unique feature that wraps over various explainability features and visuals that H2O auto-computes for a model or list of models, including the H2O AutoML object.

In this chapter, we shall explore the H2O explainability interface and how it works with the H2O AutoML object. We shall also implement a practical example to understand how to use the explainability interface in Python and R. Finally, we shall go through and understand all the various explainability features that we get as outputs.

In this chapter, we are going to cover the following main topics:

- Working with the model explainability interface
- Exploring the various explainability features

By the end of this chapter, you should have a good idea of how to interpret model performance by looking at the various performance metrics described by the model explainability interface.

Technical requirements

For this chapter, you will require the following:

- The latest version of your preferred web browser.

- An **Integrated Development Environment** (**IDE**) of your choice or a Terminal.

- All the experiments conducted in this chapter are performed on a Terminal. You are free to follow along using the same setup or you can perform the same experiments using any IDE of your choice.

All the code examples for this chapter can be found on GitHub at `https://github.com/PacktPublishing/Practical-Automated-Machine-Learning-on-H2O/tree/main/Chapter%207`.

So, let's begin by understanding how the model explainability interface works.

Working with the model explainability interface

The **model explainability interface** is a simple function that incorporates various graphs and information about the model and its workings. There are two main functions for model explainability in H2O:

- The `h2o.explain()` function, which is used to explain the model's behavior on the entire test dataset. This is also called **global explanation**.

- The `h2o.explain_row()` function, which is used to explain the model's behavior on an individual row in the test dataset. This is also called **local explanation**.

Both these functions work on either a single H2O model object, a list of H2O model objects, or the H2O AutoML object. These functions generate a list of results that consists of various graphical plots such as a **variable importance graph**, **partial dependency graph**, and a **leaderboard** if used on multiple models.

For graphs and other visual results, the `explain` object relies on visualization engines to render the graphics:

- For the R interface, H2O uses the `ggplot2` package for rendering.

- For the Python interface, H2O uses the `matplotlib` package for rendering.

With this in mind, we need to make sure that whenever we are using the explainability interface to get visual graphs, we run it in an environment that supports graph rendering. This interface won't be of much use in Terminals and other non-graphical command-line interfaces. The examples in this chapter have been run on **Jupyter Notebook**, but any environment that supports plot rendering should work fine.

The explainability function has the following parameters:

- `newdata/frame`: This parameter is used to specify the H2O test DataFrame needed to compute some of the explainability features such as the **SHAP summary** and **residual analysis**. The parameter name that's used in the R explainability interface is `newdata`, while the same in the Python explainability interface is `frame`.

- `columns`: This parameter is used to specify the columns to be considered in column-based explanations such as **individual conditional expectation plots** or **partial dependency plots**.

- `top_n_features`: This parameter is used to specify the number of columns to be considered based on the feature importance ranking for column-based explanations. The default value is 5.

  Either the `columns` parameter or the `top_n_features` parameter will be considered by the explainability function. Preference is given to the `columns` parameter, so if both the parameters are passed with values, then `top_n_features` will be ignored.

- `include_explanations`: This parameter is used to specify the explanations that you want from the explainability function's output.

- `exclude_explanations`: This parameter is used to specify the explanations that you do not want from the explainability function's output. `include_explanations` and `exclude_explanations` are mutually exclusive parameters. The available values for both parameters are as follows:

  - `leaderboard`: This value is only valid for the list of models or the AutoML object.

  - `residual_analysis`: This value is only valid for regression models.

  - `confusion_matrix`: This value is only valid for classification models.

  - `varimp`: This value stands for variable importance and is only valid for base models, not for stacked ensemble models.

  - `varimp_heatmap`: This value stands for heatmap of variable importance.

  - `model_correlation_heatmap`: This value stands for heatmap of model correlation.

  - `shap_summary`: This value stands for Shapley additive explanations.

  - `pdp`: This value stands for partial dependency plots.

  - `ice`: This value stands for individual conditional expectation plots.

- `plot_overrides`: This parameter is used to override the values for individual explanation plots. This parameter is useful if you want the top 10 features to be considered for one plot but specific columns for another:

```
list(pdp = list(top_n_features = 8))
```

- `object`: This parameter is used to specify the H2O models or the H2O AutoML object, which we will cover shortly. This parameter is specific to the R explainability interface.

Now that we know how the explainability interface works and what its various parameters are, let's understand it better with an implementation example.

We shall use **Fisher's Iris flower dataset**, which we used in *Chapter 1*, *Understanding H2O AutoML Basics*, to train models using AutoML. We will then use the explainability interface on the AutoML object to display all the explainability features it has to provide.

So, let's start by implementing it in Python.

Implementing the model explainability interface in Python

To implement the model explainability function in Python, follow these steps:

1. Import the h2o library and spin up a local H2O server:

```
library(h2o)
h2o.init(max_mem_size = "12g")
```

The explainability interface performs heavy computations behind the scenes to calculate the data needed to plot the graphs. To speed up processing, it is recommended to initialize the H2O server with as much memory as you can allocate.

2. Import the dataset using `h2o.importFile("Dataset/iris.data")`:

```
data = h2o.import_file("Dataset/iris.data")
```

3. Set which columns are the features and which columns are the labels:

```
features = data.columns
label = "C5"
```

4. Remove the label from among the features:

```
features.remove(label)
```

5. Split the DataFrame into training and testing DataFrames:

```
train_dataframe, test_dataframe = data.split_frame([0.8])
```

6. Initialize the H2O AutoML object:

```
aml = h2o.automl.H2OAutoML(max_models=10, seed = 1)
```

7. Trigger the H2O AutoML object so that it starts auto-training the models:

```
aml.train(x = features, y = label, training_frame =
train_dataframe)
```

8. Once training has finished, we can use the H2O explainability interface, h2o.explain(), on the now-trained aml object:

```
aml.explain(test_dataframe)
```

The explain function will take some time to finish computing. Once it does, you should see a big output that lists all the explainability features. The output should look as follows:

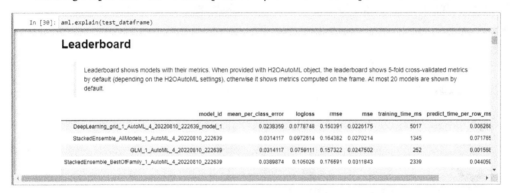

Figure 7.1 – Model explainability interface output

9. You can also use the h2o.explain_row() interface to display the model explainability features for a single row of the dataset:

```
aml.explain_row(test_dataframe, row_index=0)
```

The output of this should give you a leaderboard of the models making predictions on the first row of the dataset.

10. To get additional information about the model from an explainability point of view, you can further extend the explainability interface by using the explain_row() function on the leader model, as follows:

```
aml.leader.explain_row(test_dataframe, row_index=0)
```

The output of this should give you all the applicable graphical model explainability features for that model based on its predictions on that row.

Now that we know how to use the model explainability interface in Python, let's see how we can use this interface in the R Language.

Implementing the model explainability interface in R

Similar to how we implemented the explainability interface in Python, H2O has provisions to use the explainability interface in the R programming language as well.

To implement the model explainability function in R, follow these steps:

1. Import the h2o library and spin up a local H2O server:

```
library(h2o)
h2o.init(max_mem_size = "12g")
```

2. Import the dataset using h2o.importFile("Dataset/iris.data"):

```
data = h2o.import_file("Dataset/iris.data")
```

3. Set the C5 column as the label:

```
label <- "C5"
```

4. Split the DataFrame into training and testing DataFrames and assign them to the appropriate variables:

```
splits <- h2o.splitFrame(data, ratios = 0.8, seed = 7)
train_dataframe <- splits[[1]]
test_dataframe <- splits[[2]]
```

5. Run the H2O AutoML training:

```
aml <- h2o.automl(y = label, training_frame = train_
dataframe, max_models = 10)
```

6. Use the H2O explainability interface on the now-trained aml object:

```
explanability_object <- h2o.explain(aml, test_dataframe)
```

Once the explainability object finishes its computation, you should see a big output that lists all the explainability features.

7. Just like Python, you can also extend the model explainability interface function so that it can be run on a single row using the h2o.explain_row() function, as follows:

```
h2o.explain_row(aml, test, row_index = 1)
```

This will give you the leaderboard of models making predictions on the first row of the dataset.

8. Similarly, you can expand this explainability interface by using the `h2o.explain_row()` function on the leader model to get more advanced information about the leader model:

```
h2o.explain_row(aml@leader, test, row_index = 1)
```

In these examples, we have used the Iris flower dataset to solve a multinomial classification problem. Similarly, we can use the explainability interface on trained regression models. Some of the explainability features are only available depending on whether the trained model is a regression model or a classification model.

Now that we know how to implement the model explainability interface in Python and R, let's look deeper into the output of the interface and try to understand the various explainability features that H2O computed.

Exploring the various explainability features

The output of the explainability interface is an `H2OExplanation` object. The `H2OExplanation` object is nothing but a simple dictionary with the explainability features' names as keys. You can retrieve individual explainability features by using a feature's key name as a `dict` key on the explainability object.

If you scroll down the output of the explainability interface for the H2O AutoML object, you will notice that there are plenty of headings with explanations. Below these headings, there's a brief description of what the explainability feature is. Some have graphical diagrams, while others may have tables.

The various explainability features are as follows:

- **Leaderboard**: This feature is a leaderboard comprising all trained models and their basic metrics ranked from best performing to worst. This feature is computed only if the explainability interface is run on the H2O AutoML object or list of H2O models.

- **Confusion Matrix**: This feature is a performance metric that generates a matrix that keeps track of correct and incorrect predictions of a classification model. It is only available for classification models. For multiple models, the confusion matrix is only calculated for the leader model.

- **Residual Analysis**: This feature plots the predicted values against the residuals on the test dataset used in the explainability interface. It only analyzes the leader model based on the model ranking on the leaderboard. It is only available for regression models. For multiple models, residual analysis is performed on the leader model.

- **Variable Importance**: This feature plots the importance of variables in the dataset. It is available for all models except for stacked models. For multiple models, it is only performed on the leader model, which is not a stacked model.

- **Variable Importance Heatmap**: This feature plots a heatmap of variable importance across all the models. It is available for comparing all models except stacked models.

- **Model Correlation Heatmap**: This feature plots the correlation between the predicted values of different models. This helps group together models with similar performance. It is only available for multiple model explanations.

- **SHAP Summary of Top Tree-Based Model**: This feature plots the importance of variables in contributing to the decision-making that's done by complex tree-based models such as Random Forest and neural networks. This feature computes this plot for the top-ranking tree-based model in the leaderboard.

- **Partial Dependence Multi Plots**: This feature plots the dependency between the target feature and a certain set of features in the dataset that we consider important.

- **Individual Conditional Expectation (ICE) Plots**: This feature plots the dependency between the target feature and a certain set of features in the dataset that we consider important for each instance separately.

Comparing this to the output you got from the model explainability interface in the experiment we performed in the *Working with the model explainability interface* section, you will notice that some of the explainability features are missing from the output. This is because some of these features are only available to the type of model trained. For example, residual analysis is only available for regression models, while the experiment conducted in the *Working with the model explainability interface* section is a classification problem that trained a classification model. Hence, you won't find residual analysis in the model's explainability output.

You can perform the same experiment using a regression problem; the model explainability interface will output regression-supported explainability features.

Now that we know about the different explainability features that are available in the explanation interface, let's dive deep into them one by one to get an in-depth understanding of what they mean. We shall go through the output we got from our implementation of the explainability interface in Python and R.

In the previous chapters, we understood what the leaderboard and confusion matrix are. So, let's start with the next explanation feature: residual analysis.

Understanding residual analysis

Residual analysis is performed for **regression models**. As described in *Chapter 5, Understanding AutoML Algorithms*, in the *Understanding generalized linear models* and *Introduction to linear regression* sections, **residuals** are the difference between the values predicted by the regression model and the actual values for that same row of data. Analyzing these residual values is a great way of diagnosing any problems in your model.

A residual analysis plot is a graph where you plot the **residual values** against the **predicted values**. Another thing we learned in *Chapter 5, Understanding AutoML Algorithms*, in the *Understanding generalized linear models* and *Understanding the assumptions of linear regression* sections, is that one of the primary assumptions in **linear regression** is that the distribution of residuals is **normally distributed**.

So, accordingly, we expect our residual plot to be an amorphous collection of points. There should not be any patterns between the residual values and the predicted values.

Residual analysis can highlight the presence of **heteroscedasticity** in a trained model. Heteroscedasticity is said to have occurred if the standard deviation of the predicted values changes over different values of the features.

Consider the following diagram:

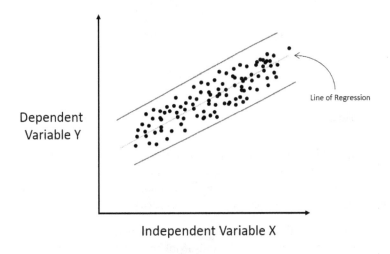

Figure 7.2 – Regression graph for a homoscedastic dataset

The preceding diagram shows a regression plot where we had some sample data that maps the relationship between X and Y. Let's fit a straight line through this data, which represents our linear model. If we calculate the residuals for every point as we go from left to right on the X-axis, we will notice that the error rate remains fairly constant throughout all the values of X. This means that all the error values lie between the parallel blue lines. Such a situation where the distribution of errors or residuals is constant throughout the independent variables is called homoscedasticity.

The opposite of homoscedasticity is **heteroscedasticity**. This is where the error rate varies over the change in the value of X. Refer to the following diagram:

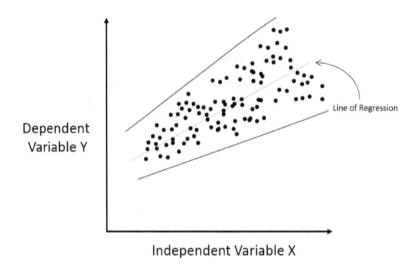

Figure 7.3 – Regression graph for a heteroscedastic dataset

As you can see, the magnitude of the errors made by the linear model increases with an increase in X. If you plot the blue error lines encompassing all the errors, then you will notice that they gradually fan out and are not parallel. This situation where the distribution of errors or residuals is not constant throughout the independent variables is called heteroscedasticity.

What heteroscedasticity tells us is that there is some sort of information that the model has not been able to capture and learn from. Heteroscedasticity also violates linear regressions' basic assumption. Thus, it can help you identify that you may need to add the missing information to your dataset to correctly train your linear model or that you may need to implement some non-linear regression algorithm to get a better-performing model.

Since residual analysis is a regression-specific model explainability feature, we cannot use the Iris dataset classification experiment that we performed in the *Working with the model explainability interface* section. Instead, we need to train a regression model and then use the model explainability interface on that model to get the residual analysis output. So, let's look at a regression problem using the Red Wine Quality dataset. You can find this dataset at `https://archive.ics.uci.edu/ml/datasets/wine+quality`.

This dataset consists of the following features:

- **fixed acidity**: This feature explains the amount of acidity that is non-volatile, meaning it does not evaporate over some time.

- **volatile acidity**: This feature explains the amount of acidity that is volatile, meaning it will evaporate over some time.

- **citric acid**: This feature explains the amount of citric acid present in the wine.

- **residual sugar**: This feature explains the amount of residual sugar present in the wine.

- **chlorides**: This feature explains the number of chlorides present in the wine.

- **free sulfur dioxide**: This feature explains the amount of free sulfur dioxide present in the wine.

- **total sulfur dioxide**: This feature explains the amount of total sulfur dioxide present in the wine.

- **density**: This feature explains the density of the wine.

- **pH**: This feature explains the pH value of the wine, with 0 being the most acidic and 14 being the most basic.

- **sulphates**: This feature explains the number of sulfates present in the wine.

- **alcohol**: This feature explains the amount of alcohol present in the wine.

- **quality**: This is the response column that notes the quality of the wine. 0 indicates that the wine is very bad, while 10 indicates that the wine is excellent.

We will run our basic H2O AutoML process of training the model and then use the model explainability interface on the trained AutoML object to get the residual analysis plot.

Now, let's observe the residual analysis plot that we get from this implementation and then see how we can retrieve the required information from the graph. Refer to the following diagram:

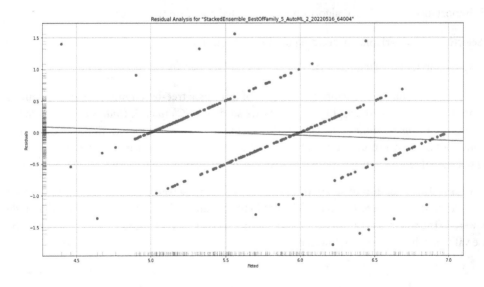

Figure 7.4 – Residual analysis graph plot for the Red Wine Quality dataset

Here, you can see the residual analysis for the stacked ensemble model, which is the leader of the AutoML trained models. On the *X*-axis, you have **Fitted**, also called predicted values, while on the *Y*-axis, you have **Residuals**.

On the left border of the *Y*-axis and below the *X*-axis you will see a **grayscale** column and row, respectively. These help you observe the distribution of those residuals across the *X* and *Y* axes.

To ensure that the distribution of residuals is normal and that the data is not heteroskedastic, you need to observe this grayscale on the *Y*-axis. A normal distribution would ideally give you a grayscale that is the darkest at the center and lightens as it moves away.

Now that you understood how to interpret the residual analysis graph, let's learn more about the next explainability feature: variable importance.

Understanding variable importance

Variable importance, also called **feature importance**, as the name suggests, explains the importance of the different variables/features in the dataset in making predictions. In any ML problem, your dataset will often have multiple variables that contribute to the characteristics of your prediction column. However, in most cases, you will often have some features that contribute more compared to others.

This understanding can help scientists and engineers remove any unwanted features that introduce noise from the dataset. This can further improve the quality of the model.

H2O calculates variable importance differently for different types of algorithms. First, let's understand how variable importance is calculated for **tree-based algorithms**.

Variable importance in tree-based algorithms is calculated based on two criteria:

- Selection of the variable for deciding on the decision tree
- Improvement in the squared error over the whole tree because of the selection

Whenever H2O is building a decision tree as a part of training a tree-based model, it will use one of the features as a node to further split the tree. As we studied in *Chapter 5, Understanding AutoML Algorithms*, in the *Understanding the Distributed Random Forest algorithm* section, we know that every node split in the decision tree aims to reduce the overall squared error. This deducted value is nothing but the difference between the squared errors of the parent node against the children node.

H2O considers this reduction in squared error in calculating the feature importance. The squared error for every node in the tree-based model leads to the variance of the response value for that node being lowered. The squared error for every node in the tree-based model leads to the variance of the response value for that node being lowered.

Thus, accordingly, the equation for calculating the squared error of the tree is as follows:

$$Squared\ Error\ (SE) = MSE * N = VAR * N$$

Here, we have the following:

- MSE means mean squared error
- N indicates the total number of observations
- VAR means variance

The equation for calculating variance is as follows:

$$VAR = \frac{1}{N} \times \sum_{i=0}^{N} (y_i - \bar{y})^2$$

Here, we have the following:

- y_i indicates the value of the observation
- $\bar{y}$ indicates the mean of all the observations
- N indicates the total number of observations

For tree-based ensemble algorithms such as the **Gradient Boosting Algorithm** (**GBM**), the decision trees are trained sequentially. Every tree is built on top of the previous tree's errors. So, the feature importance calculation is the same as how we do it for individual nodes in single decision trees.

For **Distributed Random Forest** (**DRF**), the decision trees are trained in parallel, so H2O just averages the results to calculate the feature importance.

For **XGBoost**, H2O calculates the feature importance from the gains of the loss function for individual features when building the tree.

For **deep learning**, H2O calculates the feature importance using a special method called the **Gedeon method**.

For **Generalized Linear Models** (**GLMs**), variable importance is the same as the predictor weights, also called the coefficient magnitudes. If, during training, you decide to standardize the data, then the standardized coefficients are returned.

The following diagram shows the feature importance that was calculated for our experiment on the Iris flower dataset:

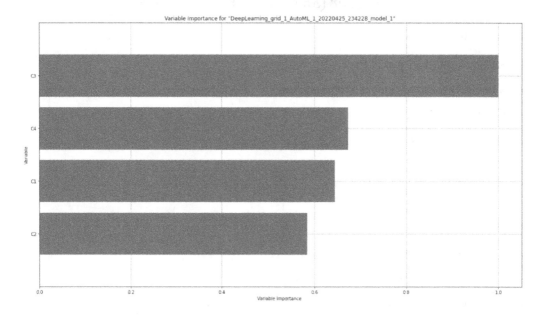

Figure 7.5 – Variable importance graph for the Iris flower dataset

The preceding diagram shows the variable importance map for a deep learning model. If you compare it with your leaderboard, you will see that the variable importance graph is plotted for the most leading model, which is not a stacked ensemble model.

On the Y-axis of the graph, you have the feature names – in our case, the **C1**, **C2**, **C3**, and **C4** columns of the Iris flower dataset. On the X-axis, you have the importance of these variables. It is possible to get the raw metric value of feature importance, but H2O displays the importance values by scaling them down between **0** and **1**, where **1** indicates the most important variable while **0** indicates the least important variable.

Since variable importance is available for both classification and regression models, you will also get a variable importance graph as an explainability feature of the Red Wine Quality regression model. The graph should look as follows:

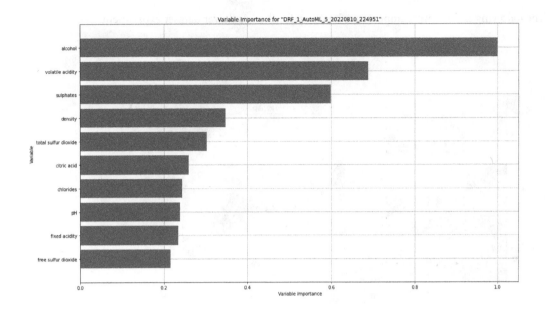

Figure 7.6 – Variable importance graph for the Red Wine Quality dataset

Now that you know how to interpret a feature importance graph, let's understand feature importance heatmaps.

Understanding feature importance heatmaps

When displaying feature importance for a specific model, it is fairly easy to represent it as a histogram or bar graph. However, we often need to compare the feature importance of various models so that we can understand which feature is deemed important by which model and how we can use this information to compare model performance. H2O AutoML will inherently train multiple models with different ML algorithms. Therefore, a comparative study of the model performance is a must and a graphical representation of feature importance can be of great help to scientists and engineers.

To represent the feature importance of all the models trained by H2O AutoML in a single graph, H2O generates a heatmap of feature importance.

A heatmap is a data visualization graph where the color of the graph is affected by the amount of density or magnitude of a specific value.

Some H2O models compute the variable importance on encoded versions of categorical columns. Different models also have different ways of encoding categorical values. So, comparing the variable importance of these categorical columns across all the models can be tricky. H2O does this comparison by summarizing the variable importance across all the features and returning a single variable importance value that represents the original categorical feature.

The following is the feature importance heatmap for the Iris flower dataset experiment:

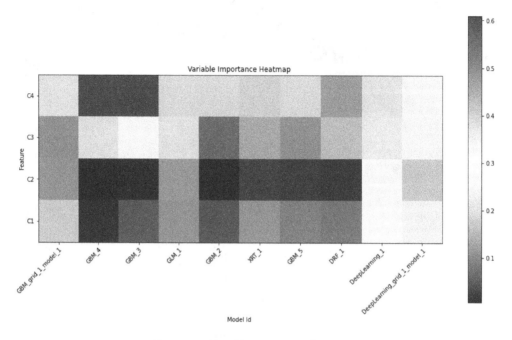

Figure 7.7 – Variable importance heatmap

Here, we can see the top 10 models on the leaderboard.

The heatmap has the **C1**, **C2**, **C3**, and **C4** features on the *Y*-axis and the model IDs on the *X*-axis. The color of the plots indicates how important the model considers the feature during its prediction. More importance equals more value, which, in turn, turns the respective plot red. The lower the importance, the lower the importance value of the feature will be; the color will become cooler and become blue.

Now that you know how to interpret feature importance heatmaps, let's learn about model correlation heatmaps.

Understanding model correlation heatmaps

Another important comparison between multiple models is **model correlation**. Model correlation can be interpreted as how similar the models are in terms of performance when you compare their prediction values.

Different models trained using the same or different ML algorithms are said to be highly correlated if the predictions made by one model are the same or similar to the predictions made by the other.

In a model correlation heatmap, H2O compares the prediction values of all the models that it trains and compares them to one another.

The following is the model correlation heatmap graph we got from our experiment on the Iris flower dataset:

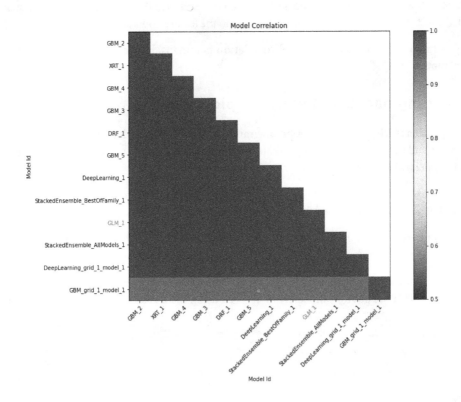

Figure 7.8 – Model correlation heatmap for the Iris flower dataset

> **Tip**
> To understand this explainability feature graph, kindly refer to the *Model Correlation* section in the output you got after executing the `explain()` function in your code.

On the *X* and *Y* axes, we have the model IDs. Their cross-section on the graph indicates the correlation value between them. You will notice that the heat points on the graph within the *X* and *Y* axes have the same model ID, which will always be 1; therefore, the plot will always be red. This is correct as, technically, it's the same model and when you compare the prediction values of a model with itself, there will be 100% correlation.

To get a better idea of the correlation between different models, you can refer to these heat values. Dark red points indicate high correlation, while those with cool blue values indicate low correlation. Models highlighted in red are interpretable models such as GLMs.

You may notice that since the model correlation heatmap supports stacked ensemble models and feature importance heatmaps don't, if you ignore the stacked ensemble models in the model correlation heatmap (*Figure 7.8*), the rest of the models are the same as the ones in the feature importance heatmap (*Figure 7.7*).

Now that you know how to interpret model correlation heatmaps, let's learn more about partial dependency plots.

Understanding partial dependency plots

A **Partial Dependence Plot** (**PDP**) is a graph diagram that shows you the dependency between the predicted values and the set of input features that we are interested in while marginalizing the values of features in which we are not interested.

Another way of understanding a PDP is that it represents a function of input features that we are interested in that gives us the expected predicted values as output.

A PDP is a very interesting graph that is useful in showing and explaining the model training results to members of the organization that are not so skilled in the domain of data science.

First, let's understand how to interpret a DPD before learning how it is calculated. The following diagram shows the PDP we got for our experiment when using the Iris flower dataset:

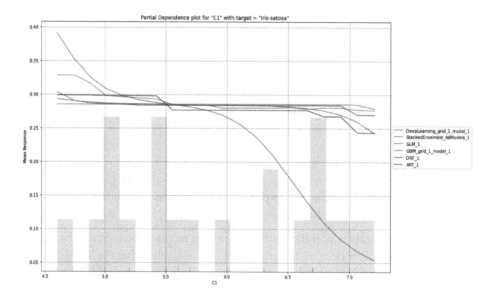

Figure 7.9 – PDP for the C1 column with Iris-setosa as the target

> **Tip**
>
> To understand this explainability feature graph, kindly refer to the *Partial Dependency Plots* section in the output you got after executing the `explain()` function in your code.

The PDP plot is a graph that shows you the marginal effects of a feature on the response values. On the X-axis of the graph, you have the selected feature and its range of values. On the Y-axis, you have the mean response values for the target value. The PDP plot aims to tell the viewer what the mean response value predicted by the model for a given value of the selected feature is.

In *Figure 7.9*, the PDP graph is plotted for the **C1** column for the target value, which is **Iris-setosa**. On the X-axis, we have the **C1** column, which stands for the sepal length of the flower in centimeters. The range of these values ranges from the minimum value present in the dataset to the maximum value. On the Y-axis, we have the mean response values. For this experiment, the mean response values are probabilities that the flower is an Iris-setosa, which is the selected target value of the plot. The colorful lines on the graph indicate the mean response values predicted by the different models trained by H2O AutoML for the range of **C1** values.

Looking at this graph gives us a good idea of how the response value is dependent on the single feature, **C1**, for every individual model. We can see that so long as the sepal length lies between 4.5 to 6.5 centimeters, most of the models show an approximate probability that there is a 35% chance that the flower is of the Iris-setosa class.

Similarly, in the following graph, we have plotted the PDP graph for the **C1** column, only this time the target response column is **Iris-versicolor**:

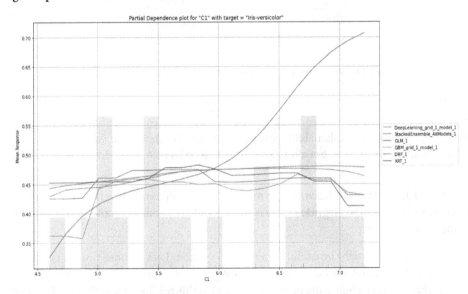

Figure 7.10 – PDP for the C1 column with Iris-versicolor as the target

Here, we can see that so long the values of **C1** are between 4.5 to 6.5, there is around a 27% to 40% chance that the flower is of the Iris-versicolor class. Now, let's look at the following PDP plot for **C1** for the third target value, **Iris-virginica**:

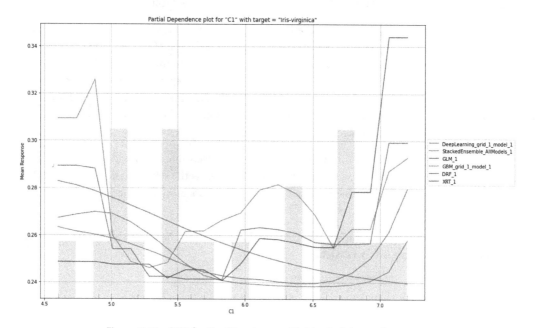

Figure 7.11 – PDP for the C1 column with Iris-virginica as the target

You will notice that, for **Iris-virginica**, all the models predict differently for the same values of **C1**. This could mean that the **Iris-virginica** class is not strongly dependent on the sepal length of the flower – that is, the **C1** value.

Another case where PDP might be useful is in model selection. Let's assume you are certain that a specific feature in your dataset will be contributing greatly to the response value and you train multiple models on it. Then, you can choose the model that best suits this relationship as that model will make the most realistically accurate predictions.

Now, let's try to understand how the PDP plot is generated and how H2O computes these plot values.

The PDP plot data can be calculated as follows:

1. Choose a feature and target value to plot the dependency.
2. Bootstrap a dataset from the validation dataset, where the value of the selected feature is set to the minimum value present in the validation dataset for all rows.
3. Pass this bootstrapped dataset to one of the models trained by H2O AutoML and calculate the mean of the prediction values it got for all rows.
4. Plot this value on the PDP graph for that model.
5. Repeat *steps 3* and *4* for the remaining models.
6. Repeat *step 2*, but this time, increment the value of the selected feature to the next value present in the validation dataset. Then, repeat the remaining steps.

You will do this for all the feature values present in the validation dataset and plot them on the results for all the models on the same PDP graph.

Once finished, you will repeat the same process for different combinations of the feature and target response values.

H2O will make multiple PDP plots for all the combinations of features and response values. The following is a PDP plot where the selected feature is **C2** and the selected target value is **Iris-setosa**:

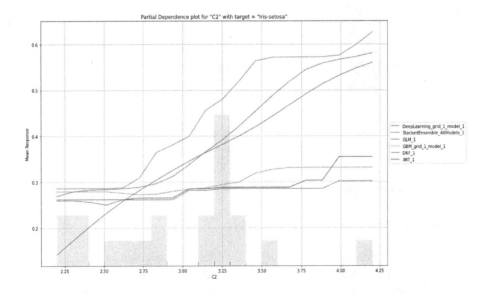

Figure 7.12 – PDP for the C2 column with Iris-setosa as the target

Similarly, it created different combinations of the PDP plot for the **C3** and **C4** features. The following is a PDP plot where the selected feature is **C3** and the selected target value is **Iris-versicolor**:

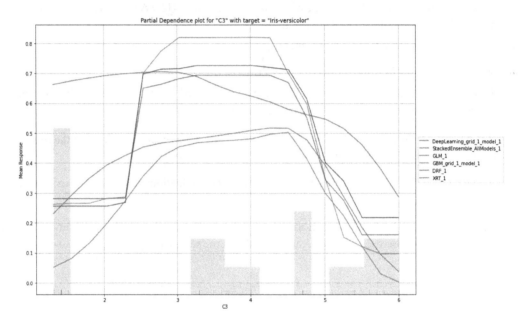

Figure 7.13 – PDP for the C3 column with Iris-versicolor as the target

Now that you know how to interpret feature importance heatmaps, let's learn about SHAP summary plots.

Understanding SHAP summary plots

For sophisticated problems, tree-based models can become difficult to understand. Complex tree models can be very large and complicated to understand. The **SHAP summary plot** is a simplified graph of the tree-based model that gives you a summarized view of the model's complexity and how it behaves.

SHAP stands for **Shapley Additive Explanations**. SHAP is a model explainability feature that takes an approach from game theory to explain the output of an ML model. The SHAP summary plot shows you the contribution of the features toward predicting values, similar to PDPs.

Let's try to interpret a SHAP value from an example. The following is the SHAP summary we get from the Red Wine Quality dataset:

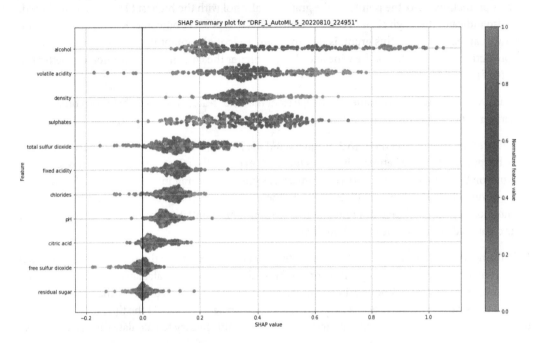

Figure 7.14 – SHAP summary plot for the Red Wine Quality dataset

> **Tip**
>
> To better understand this explainability feature graph, kindly refer to the *SHAP Summary* section in the output you get after executing the `explain()` function on your regression model.

On the right-hand side, you can see a bluish-red bar. This bar represents the normalized value of the wine quality in color. The redder the color, the better the quality; the bluer the color, the poorer the wine quality. In binomial problems, the color will be a stark contrast between red and blue. However, in regression problems, like in our example, we can have a whole spectrum of colors, indicating the range of possible numerical values.

On the *Y*-axis, you have the features from the dataset. They are in descending order from top to bottom based on the feature's importance. In our example, the alcohol content is the most important feature in the dataset; it contributes more to the final prediction value.

On the *X*-axis, you have the **SHAP value**. The SHAP value denotes how the feature helps the model toward the expected outcome. The more positive the SHAP value, the more the feature contributes to the outcome.

Let's take the example of alcohol from the SHAP summary. Based on this, we can see that alcohol has the highest SHAP value among the rest of the features. Thus, alcohol contributes greatly to the model's prediction. Also, the points on the graph for alcohol with the highest SHAP value are in red. This also indicates that high alcohol content contributes to a positive outcome. Keeping this in mind, what we can extract from this graph is that the feature alcohol content plays an important part in the prediction of the quality of the wine and that the higher the content of the alcohol, the better the quality of the wine.

Similarly, you can interpret the same knowledge from the other features. This can help you compare and understand which features are important and how they contribute to the final prediction of the model.

One interesting question on the SHAP summary and the PDP is, what is the difference between them? Well, the prime difference between the two is that PDP explains the effect of replacing only one feature at a time on the output, while the SHAP summary considers the overall interaction of that feature with other features in the dataset. So, PDP works on the assumption that your features are independent of one another, while SHAP takes into account the combined contributions of different features and their combined effects on the overall prediction.

Calculating **SHAP values** is a complex process that is derived from game theory. If you are interested in expanding your knowledge of game theory and how SHAP values are calculated, feel free to explore them at your own pace. A good starting point for understanding SHAP is to follow the explanations at `https://shap.readthedocs.io/en/latest/index.html`. At the time of writing, H2O acts as a wrapper for the SHAP library and internally uses this library to calculate the SHAP values.

Now that we know how to interpret a SHAP summary plot, let's learn about explainability feature, **Individual Conditional Expectation** (ICE) plots.

Understanding individual conditional expectation plots

An **ICE** plot is a graph that displays a line for every instance of an observation that shows how the prediction for the given observation changes when the value of a feature changes.

ICE plots are similar to PDP graphs. PDP focuses on the overall average effect of a change in a feature on the prediction outcome, while ICE plots focus on the dependency of the outcome on individual instances of the feature value. If you average the ICE plot values, you should get a PDP.

The way to compute ICE plots is very simple, as shown in the following screenshot:

	Feature 1	Feature 2	Feature 3	Feature 4	...	Target
Observation 1	5	57	8	4	...	1
Observation 2	2	11	47	67	...	1
Observation 3	6	43	84	8	...	0
Observation 4	7	3	46	457	...	1
...	...	...	...	...	...	...
Observation N	3	5	27	37	...	1

Figure 7.15 – Sample dataset for ICE graph plots highlighting Observation 1

Once your model has been trained, you must perform the following steps to calculate the ICE plots:

1. Consider the first observation – in our example, **Observation 1** – and plot the relationship between **Feature 1** and the respective **Target** value.

2. Keeping the values in **Feature 1** constant, create a bootstrapped dataset while replacing all the other feature values with those seen in **Observation 1** in the original dataset; mark all other observations as **Observation 1**.

3. Calculate the **Target** value of the observations using your trained model.

    Refer to the following screenshot for the bootstrapped dataset:

	Feature 1	Feature 2	Feature 3	Feature 4	...	Target
Observation 1	5	57	8	4	...	1
Observation 1	2	57	8	4	...	0.34
Observation 1	6	57	8	4	...	0.72
Observation 1	7	57	8	4	...	0.21
...	...	...	...	...	...	...
Observation 1	3	57	8	4	...	0.71

Figure 7.16 – Bootstrapped dataset for Observation 1 for Feature 1

4. Repeat the same for the next observation. Consider the second observation – in our example, **Observation 2** – and plot the relationship between **Feature 1** and the respective **Target** value:

	Feature 1	Feature 2	Feature 3	Feature 4	...	Target
Observation 1	5	57	8	4	...	1
Observation 2	2	11	47	67	...	1
Observation 3	6	43	84	8	...	0
Observation 4	7	3	46	457	...	1
...	...	...	...	...	...	...
Observation N	3	5	27	37	...	1

Figure 7.17 – Sample dataset for an ICE plot highlighting Observation 2

5. Keep the values in **Feature 1** constant and create a bootstrapped dataset; then, calculate the **Target** values using the trained model. Refer to the following resultant bootstrapped dataset:

	Feature 1	Feature 2	Feature 3	Feature 4	...	Target
Observation 2	5	11	47	67	...	0.34
Observation 2	2	11	47	67	...	1
Observation 2	6	11	47	67	...	0.77
Observation 2	7	11	47	67	...	0.84
...	...	...	...	...	...	...
Observation 2	3	11	47	67	...	0.75

Figure 7.18 – Bootstrapped dataset for Observation 2 for Feature 1

6. We repeat this process for all observations against all features.

7. The results that are observed from these bootstrapped datasets are plotted on the individual ICE plots per feature.

Let's see how we can interpret the ICE plot and extract observable information out of the graph. Refer to the following screenshot, which shows the ICE plot we get after running the model explainability interface on the AutoML object that was trained on the Red Wine Quality dataset:

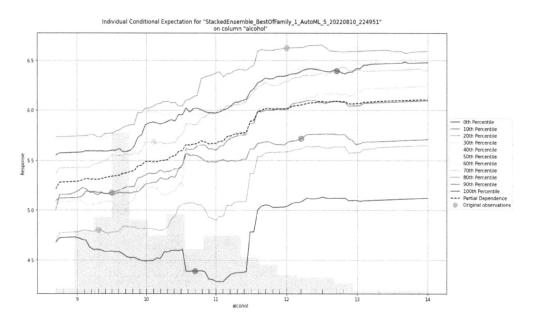

Figure 7.19 – ICE plot for the Red Wine Quality dataset

As the heading states, this is an ICE plot on the alcohol feature column of the dataset for a stacked ensemble model trained by H2O AutoML. Keep in mind that this model is the leader of the list of models trained by AutoML. ICE plots are only plotted for the leader of the dataset. You can also observe the ICE plots of the other models by extracting the models using their model IDs and then running the `ice_plot()` function on them. Refer to the following code example:

```
model = h2o.get_model("XRT_1_AutoML_2_20220516_64004")
model.ice_plot(test_dataframe, "alcohol")
```

On the *X*-axis of the graph, you have the range of values for the alcohol feature. On the *Y*-axis, you have the range of values of the predicted outcomes – that is, the quality of the wine.

On the left-hand side of the graph, you can see the legends stating different types of lines and their percentiles. The ICE plot plots the effects for each decile. So, technically, when plotting the ICE plot, you compute a line for every observation. However, in a dataset that contains thousands if not millions of rows of data, you will end up with an equal number of lines on the plot. This will make the ICE plot messy. That is why to better observe this data, you must aggregate the lines together to the nearest decile and plot a single line for every percentile partition.

The dotted black line is the average of all these other percentile lines and is nothing but the PDP line for that feature.

Now that you know how to interpret an ICE plot, let's look at learning curve plots.

Understanding learning curve plots

The **learning curve plot** is one of the most used plots by data scientists to observe the learning rate of a model. The **learning curve** shows how your model learns from the dataset and the efficiency with which it does the learning.

When working on an ML problem, an important question that often needs to be answered is, *how much data do we need to train the most accurate model?* A learning curve plot can help you understand how increasing the dataset affects your overall model performance.

Using this information, you can decide if increasing the size of the dataset can result in better model performance or if you need to work on your model training to improve your model's performance.

Let's observe the learning curve plot we got from our experiment on the Red Wine Quality dataset for the XRT model trained by AutoML:

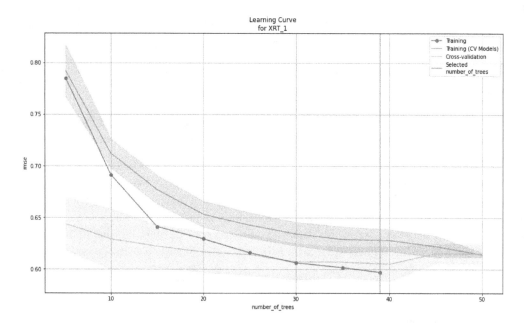

Figure 7.20 – Learning curve plot for the XRT model on the Red Wine Quality dataset

On the *X*-axis of the graph, you have the number of trees created by the XRT algorithm. As you can see, the algorithm created around 40 to 50 trees in total. On the *Y*-axis, you have the performance metric, RMSE, which is calculated at every stage during the model training as the algorithm creates the trees.

As shown in the preceding screenshot, the RMSE metric decreases as the algorithm creates more trees. Eventually, the rate at which the RMSE lowers decreases over a certain number of trees created. Any trees created over this number do not contribute to the overall improvement in the model's performance. Thus, the learning rate eventually decreases over the increase in several trees.

The lines on the graph depict the various datasets that were used by the algorithm during training and the respective RMSE during every instance of creating the trees.

At the time of writing, as of H2O version *3.36.1*, the learning curve plot is not part of the default model explainability interface. To plot the learning curve, you must plot it using the following function on the respective model object:

```
model = h2o.get_model("GLM_1_AutoML_2_20220516_64004")
model.learning_curve_plot()
```

The learning curve plot is different for different algorithms. The following screenshot shows a learning plot for a GLM model trained by AutoML on the same dataset:

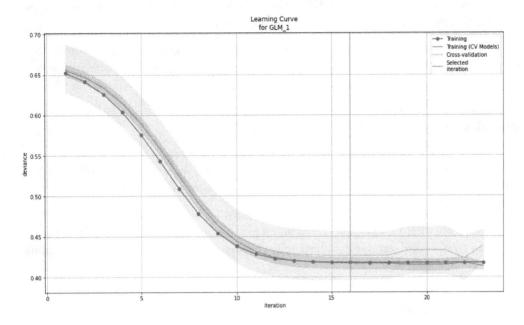

Figure 7.21 – Learning curve plot for the GLM model on the Red Wine Quality dataset

As you can see, instead of the number of trees on the X-axis, we now have iterations. The number of trees is relevant for tree-based algorithms such as XRT and DRF, but linear models such as GLM running on the linear algorithm makes more sense to aid with learning. On the Y-axis, you have deviance instead of RMSE as deviance is more suitable for measuring the performance of a linear model.

The learning curve is different for different types of algorithms, including stacked ensemble models. Feel free to explore the different variations of the learning curve for different algorithms. H2O already takes care of selecting the appropriate performance metric and the steps in learning, depending on the algorithms, so you don't have to worry about whether you chose the right metric to measure the learning rate or not.

Summary

In this chapter, we focused on understanding the model explainability interface provided by H2O. First, we understood how the explainability interface provides different explainability features that help users get detailed information about the models trained. Then, we learned how to implement this functionality on models trained by H2O's AutoML in both Python and R.

Once we were comfortable with its implementation, we started exploring and understanding the various explainability graphs displayed by the explainability interface's output, starting with residual analysis. We observed how residual analysis helps highlight heteroscedasticity in the dataset and how it helps you identify if there is any missing information in your dataset.

Then, we explored variable importance and how it helps you identify important features in the dataset. Building on top of this, we learned how feature importance heatmaps can help you observe feature importance among all the models trained by AutoML.

Then, we discovered how model correlation heatmaps can be interpreted and how they help us identify models with similar prediction behavior from a list of models.

Later, we learned about PDP graphs and how they express the dependency of the overall outcome over the individual features of the dataset. With this knowledge in mind, we explored the SHAP summary and ICE plots, where we understood the two graphs and how each focuses on different aspects of outcome dependency on individual features.

Finally, we explored what a learning plot is and how it helps us understand how the model improves in performance, also called learning, over the number of observations, iterations, or trees, depending on the type of algorithms used to train the model.

In the next chapter, we shall use all the knowledge we've learned from the last few chapters and explore the other advanced parameters that are available when using H2O's AutoML feature.

Part 3
H2O AutoML Advanced Implementation and Productization

This part will help you understand H2O AutoML's advanced features and parameters used to customize certain characteristics of AutoML to suit specialized needs. This will help you get the desired personalized results that generalized machine learning fails to provide. It will also explain the various ways that H2O AutoML can be used with different types of technologies, and you will understand how you can deploy your machine learning models into production, and commercially use them to meet business needs.

This section comprises the following chapters:

- *Chapter 8, Exploring Optional Parameters for H2O AutoML*
- *Chapter 9, Exploring Miscellaneous Features in H2O AutoML*
- *Chapter 10, Working with Plain Old Java Objects (POJOs)*
- *Chapter 11, Working with Model Object Optimized (MOJO)*
- *Chapter 12, Working with H2O AutoML and Apache Spark*
- *Chapter 13, Using H2O AutoML with Other Technologies*

Exploring Optional Parameters for H2O AutoML

As we explored in *Chapter 2, Working with H2O Flow (H2O's Web UI)*, when training models using H2O AutoML, we had plenty of parameters to select. All these parameters gave us the capability to control how H2O AutoML should train our models. This control helps us get the best possible use of AutoML based on our requirements. Most of the parameters we explored were pretty straightforward to understand. However, there were some parameters whose purpose and effects were slightly complex to be understood at the very start of this book.

In this chapter, we shall explore these parameters by learning about the **Machine Learning** (**ML**) concepts behind them, and then understand how we can use them in an AutoML setting.

By the end of this chapter, you will not only be educated in some of the advanced ML concepts, but you will also be able to implement them using the parametric provisions made in H2O AutoML.

In this chapter, we will cover the following topics:

- Experimenting with parameters that support imbalanced classes
- Experimenting with parameters that support early stopping
- Experimenting with parameters that support cross-validation

We will start by understanding what imbalanced classes are.

Technical requirements

You will require the following to complete this chapter:

- The latest version of your preferred web browser.
- The H2O software installed on your system. Refer to *Chapter 1, Understanding H2O AutoML Basics*, for instructions on how to install H2O on your system.

> **Tip**
>
> All the H2O AutoML function parameters shown in this chapter are shown using H2O Flow to keep things simple. The equivalent parameters are also available in the Python and R programming languages for software engineers to code into their services. You can find these details at `https://docs.h2o.ai/h2o/latest-stable/h2o-docs/parameters.html`.

Experimenting with parameters that support imbalanced classes

One common problem you will often face in the field of ML is classifying rare events. Consider the case of large earthquakes. Large earthquakes of magnitude 7 and higher occur about once every year. If you had a dataset containing the Earth's tectonic activity of each day since the last decade with the response column containing whether or not an earthquake occurred, then you would have approximately 3,650 rows of data; that is, one row for each day in the decade, with around 8-12 rows showing large earthquakes. That is less than a 0.3% chance that this event will occur. 99.7% of the time, there will be no large earthquakes. This dataset, where the number of large earthquake events is so small, is called an **imbalanced dataset**.

The problem with the imbalanced dataset is that even if you write a simple `if-else` function that marks all tectonic events as not earthquakes and call this a model, it will still show the accuracy as 99.7% accuracy since the majority of the events are not earthquake-causing. However, in actuality, this so-called model is very bad as it is not correctly informing you whether it is an earthquake or not.

Such imbalance in the **target class** creates a lot of issues when training ML models. The ML models are more likely to assume that these events are so rare that they will never occur and will not learn the distinction between those events.

However, there are ways to tackle this issue. One way is to undersample the majority class and the other way is to oversample the minority class. We shall learn more about these techniques in the upcoming sections.

Understanding undersampling the majority class

In the scenario of predicting the occurrence of earthquakes, the dataset contains a large number of events that have been identified as *not-earthquake*. This event is known as the majority class. The few events that mark the activity as an *earthquake* are known as the minority class.

Let's see how **undersampling the majority class** can solve the problems caused by an imbalance in the classes. Consider the following diagram:

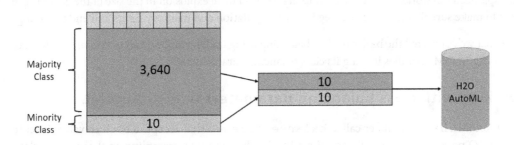

Figure 8.1 – Undersampling an imbalanced dataset

Let's assume you have 3,640 data samples of tectonic activity that indicate no earthquakes happened and only 10 samples that indicate earthquakes happened. In this case, to tackle this imbalance issue, you must create a bootstrapped dataset containing all 10 samples of the minority class, and 10 samples of the majority class chosen at random from the 3,640 data samples. Then, you can feed this new dataset to H2O AutoML for training. In this case, we have undersampled the majority class and equalized the *earthquake* and *not-earthquake* data samples before training the model.

The drawback of this approach is that we end up tossing away huge amounts of data, and the model won't be able to learn a lot from the reduced data.

Understanding oversampling the minority class

The second approach to tackling the imbalanced dataset issue is to **oversample the minority class**. One obvious way is to duplicate the minority class data samples and append them to the dataset so that the number of data samples between the majority and minority classes is equal. Refer to the following diagram for a better understanding:

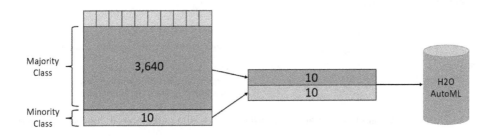

Figure 8.2 – Oversampling an imbalanced dataset

In the preceding diagram, you can see that we replicated the minority class data samples and appended them to the dataset so that we ended up with 3,640 rows for each of the classes.

This approach can work; however, oversampling will lead to an explosion in the size of the dataset. You need to make sure that it does not exceed your computation and memory limits and end up failing.

Now that we've covered the basics of class balancing using undersampling and oversampling, let's see how H2O AutoML handles it using its class balancing parameters.

Working with class balancing parameters in H2O AutoML

H2O AutoML has a parameter called `balance_classes` that accepts a boolean value. If set to *True*, H2O performs oversampling on the minority class and undersampling on the majority class. The balancing is performed in such a way that eventually, each class contains the same number of data samples.

Both undersampling and oversampling of the respective classes is done randomly. Additionally, oversampling of the minority class is done with replacement. This means that data samples from the minority class can be chosen and added to the new training dataset multiple times and can be repeated.

H2O AutoML has the following parameters that support class balancing functionality:

- `balance_classes`: This parameter accepts a boolean value. It is *False* by default, but if you want to perform class balancing on your dataset before feeding it to H2O AutoML for training, then you can set the boolean value to *True*.

  In H2O Flow, you get a checkbox besides the parameter. Refer to the following screenshot:

  balance_classes ☑   Balance training data class counts via over/under-sampling (for imbalanced data).

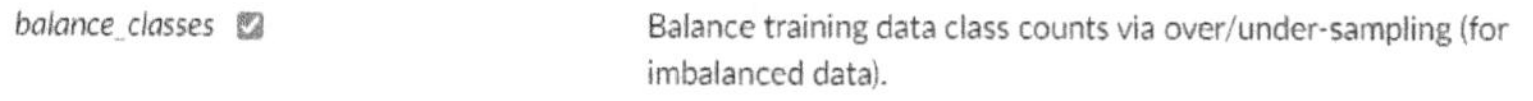

  Figure 8.3 – The balance_classes checkbox in H2O Flow

  Checking it makes the `class_sampling_factors` and `max_after_balance_size` parameters available in the **EXPERT** section of the **Run AutoML** parameters, as shown in the following screenshot:

  EXPERT

  class_sampling_factors   Desired over/under-sampling ratios per class (in lexicographic order). If not specified, sampling factors will be automatically computed to obtain class balance during training. Requires balance_classes.

  max_after_balance_size   5   Maximum relative size of the training data after balancing class counts (defaults to 5.0 and can be less than 1.0). Requires balance_classes.

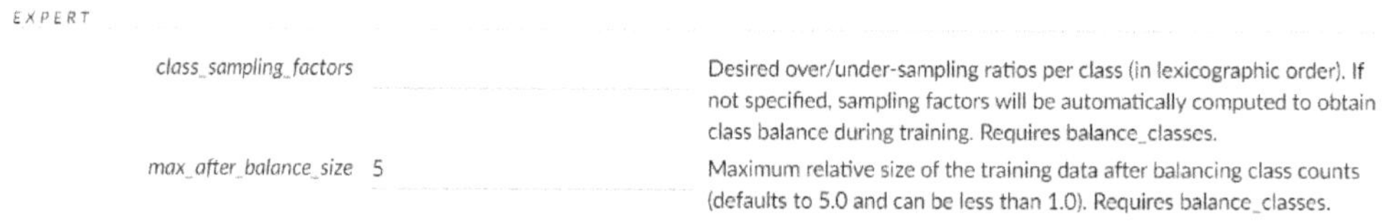

Figure 8.4 – The class_sampling_factors and max_after_balance_size parameters in the EXPERT section

- `class_sampling_factors`: This parameter requires `balance_classes` to be *True*. This parameter takes a list of float values as input that will represent the sampling rate for that class. A sampling rate of value *1.0* for a given class will not change its sample rate during class balancing. A sampling rate of *0.5* will halve the sample rate of a class during class balancing while a sampling rate of *2.0* will double it.

- `max_after_balance_size`: This parameter requires `balance_classes` to be *True* and specifies the maximum relative size of the training dataset after balancing. This parameter accepts a `float` value as input, which would limit the size your training dataset can grow to. The default value is *5.0*, which indicates that the training dataset will grow a maximum of *5* times its size. This value can also be less than *1.0*.

In the Python programming language, you can set these parameters as follows:

```
aml = h2o.automl.H2OAutoML(balance_classes = True, class_
sampling_factors =[0.3, 2.0], max_after_balance_size=0.95, seed
= 123)
aml.train(x = features, y = label, training_frame = train_
dataframe)
```

Similarly, in the R programming language, you can set these parameters as follows:

```
aml <- h2o.automl(x = features, y = label, training_frame =
train_dataframe, seed = 123, balance_classes = TRUE, class_
sampling_factors = c(0.3, 2.0), max_after_balance_size=0.95)
```

To perform class balancing when training models using AutoML, you can set the `balance_classes` parameter to true in the H2O AutoML estimator object. In that same object, you can specify your `class_sampling_factors` and `max_after_balance_size` parameters. Then, you can use this initialized AutoML estimator object to trigger AutoML on your training dataset.

Now that you understand how we can tackle the class imbalance issue using the `balance_classes`, `class_sampling_factors`, and `max_after_balance_size` parameters, let's understand the next optional parameters in AutoML – that is, stopping criteria.

Experimenting with parameters that support early stopping

Overfitting models is one of the common issues often faced when trying to solve an ML problem. Overfitting is said to have occurred when the ML model tries to adapt to your training set too much, so much so that it is only able to make predictions on values that it has seen before in the training set and is unable to make a generalized prediction on unseen data.

Overfitting occurs due to a variety of reasons, one of them being that the model learns so much from the dataset that it even incorporates and learns the noise in the dataset. This learning negatively impacts predictions on new data that may not have that noise. So, how do we tackle this issue and prevent the model from overfitting? Stop the model early before it learns the noise.

In the following sub-sections, we shall understand what early stopping is and how it is done. Then, we will learn how the early stopping parameters offered by H2O AutoML work.

Understanding early stopping

Early stopping is a form of **regularization** that stops a model's training once it has achieved a satisfactory understanding of the data and further prevents it from overfitting. Early stopping aims to observe the model's performance as it improves using an appropriate performance metric and stop the model's training once deterioration is observed due to overfitting.

When training a model using algorithms that use iterative optimization to minimize the loss function, the training dataset is passed through the algorithm during each iteration. Observations and understandings that pass are then used during the next iteration. This iteration of passing the training dataset through the algorithm is called an **epoch**.

For early stopping, at the end of every epoch, we can calculate the performance of the model and note down the metric value. Comparing these values during every iteration helps us understand whether the model is improving its performance after every epoch or whether it is learning noise and losing performance. We can monitor this and stop the model training at the epoch where we start seeing a decrease in performance. Refer to the following diagram to gain a better understanding of early stopping:

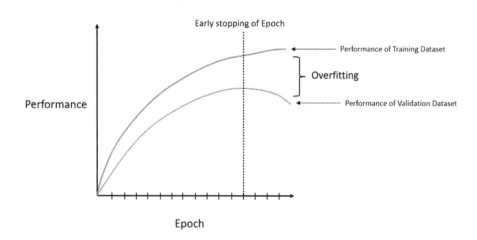

Figure 8.5 – Early stopping to avoid model overfitting

In the preceding diagram, on the *Y* axis, we have the **Performance** value of the model. On the *X*-axis, we have the **Epoch** value. So, as time goes on and we iterate through the number of epochs, we see that the performance of the model on the training set and the validation set continues to increase. But after a certain point, the performance of the model on the validation dataset starts decreasing, while the performance of the training dataset continues to increase. This is where overfitting starts. The model learns too much from the training dataset and starts incorporating noise into its learning. This might show high performance on the training dataset, but the model fails to generalize the predictions. This leads to bad predictions on unseen data, such as the ones in the validation dataset.

So, the best thing to do is to stop the model at the exact point where the performance of the model is highest for both the training and validation dataset.

Now that we have a basic understanding of how early stopping of model training works, let's learn how we can perform it using the early stopping parameter offered by the H2O AutoML function.

Working with early stopping parameters in H2O AutoML

H2O AutoML has provisions for you to implement and control the early stopping of your models that it will auto train for you.

You can use the following parameters to implement early stopping:

- `stopping_rounds`: This parameter indicates the number of training rounds over which if the stopping metric fails to improve, we stop the model training.

- `stopping_metric`: This parameter is used to select the performance metric to consider when early stopping. It is available if `stopping_rounds` is set and is greater than *0*. We studied performance metrics in *Chapter 6, Understanding H2O AutoML Leaderboard and Other Performance Metrics*, so kindly refer to that chapter if you wish to revise how the different metrics measure performance. The available options for this parameter are as follows:

  - `AUTO`: This is the default value and further defaults to the following values, depending on the type of ML problem:

    - `logloss`: The default stopping metric for classification problems.

    - `deviance`: The default stopping metric for regression problems. This stands for mean residual deviance.

    - `anomaly_score`: The default stopping metric for Isolation Forest models, which are a type of ensemble model.

    - `anomaly_score`: The default stopping metric for Isolation Forest models (ensemble models). It is the measure of normality of an observation equivalent to the number of splits in a decision tree needed to isolate a point in a given tree where that point is at max depth.

- deviance: This stands for mean residual deviance. This value tells us how well the label value can be predicted by a model based on the number of features in the dataset.

- logloss: Log loss is a metric that is a way of measuring the performance of a classification model that outputs classification results in the form of probability values.

- MSE (**Mean Squared Error**): This is a metric that measures the mean of the squares of errors of the predicted value against the actual value.

- RMSE (**Root Mean Squared Error**): This is a metric that calculates the root value of the MSE.

- MAE (**Mean Absolute Error**): This is a metric that calculates the average magnitude of errors in each set of observations.

- RMSLE (**Root Mean Squared Logarithmic Error**): This is a metric that calculates the RMSE of the log-transformed observed and log-transformed actual values.

- AUC (**Area Under the ROC Curve**): AUC-ROC is a metric that helps us compare which classification algorithm performed better, depending on whose ROC curve covers the most area.

- AUCPR (**Area Under the Precision-Recall Curve**): AUCPR is similar to the AUC-ROC curve, with the only difference being that the PR curve is a function that uses precision on the Y axis and recall on the X axis.

- lift_top_group: This parameter configures AutoML in such a way that the model being trained must improve its lift within the top 1% of the training data. Lift is nothing but the measure of performance of a model in making accurate predictions, compared to a model that randomly makes predictions. The top 1% of the dataset are the observations with the highest predicted values.

- misclassification: This metric is used to measure the fraction of the predictions that were incorrectly predicted without distinguishing between positive and negative predictions.

- mean_per_class_error: This is a metric that calculates the average of all errors per class in a dataset containing multiple classes.

- custom: This parameter is used to set any custom metric as the stopping metric during AutoML training. The custom metric should be of the behavior *less is better*, meaning the lower the value of the custom metric, the better the performance of the model. The lower bound value of the custom metric is assumed to be 0.

- custom_increasing: This parameter is for custom performance metrics that have the behavior as *more is better*, meaning the higher the value of these metrics, the better the model performance. At the time of writing, this parameter is only supported in the Python client for GBM and DRF.

- `stopping_tolerance`: This parameter indicates the tolerance value by which the model's performance metric must improve before stopping the model training. It is available if `stopping_rounds` is set and is greater than *0*. The default stopping tolerance for AutoML is *0.001* if the dataset contains at least 1 million rows; otherwise, the value is determined by the size of the dataset and the amount of non-NA data in the dataset, which leads to a value greater than *0.001*

In H2O Flow, these parameters are available in the **ADVANCED** section of the **Run AutoML** parameters, as shown in the following screenshot:

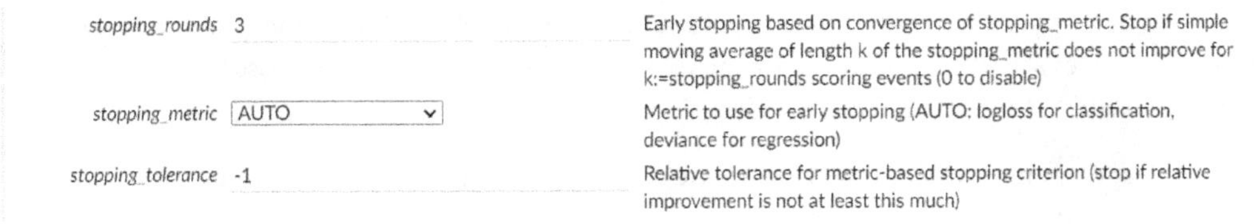

stopping_rounds 3		Early stopping based on convergence of stopping_metric. Stop if simple moving average of length k of the stopping_metric does not improve for k:=stopping_rounds scoring events (0 to disable)
stopping_metric AUTO ⌄		Metric to use for early stopping (AUTO: logloss for classification, deviance for regression)
stopping_tolerance -1		Relative tolerance for metric-based stopping criterion (stop if relative improvement is not at least this much)

Figure 8.6 – Early stopping parameters in H2O Flow

In the Python programming language, you can set these parameters as follows:

```
aml = h2o.automl.H2OAutoML(stopping_metric = "mse",
stopping_rounds = 5, stopping_tolerance = 0.001)
aml.train(x = features, y = label, training_frame =
train_dataframe)
```

In the R programming language, you can set these parameters as follows:

```
aml <- h2o.automl(x = features, y = label, training_frame
= train_dataframe, seed = 123, stopping_metric = "mse",
stopping_rounds = 5, stopping_tolerance = 0.001)
```

To better understand how AutoML will stop a model's training early, consider the same Python and R example values. We have `stopping_metric` as **mse**, `stopping_rounds` as **5**, and `stopping_tolerance` as **0.001**.

When implementing early stopping, H2O will calculate the moving average of the last **6** stopping rounds, where the average of the very first round is used as a reference for the next rounds. If the ratio between the best moving average and the reference moving average is greater than or equal to a `stopping_tolerance` of *0.001*, then H2O will stop the model training. For performance metrics that have the *more is better* behavior, the ratio between the best moving average and reference moving average should be less than or equal to the stopping tolerance.

Now that we understand how to stop model training early using the `stopping_rounds`, `stopping_metrics`, and `stopping tolerance` parameters, let's understand the next optional parameter in AutoML – that is, cross-validation.

Experimenting with parameters that support cross-validation

When performing model training on a dataset, we usually perform a train-test split on the dataset. Let's assume we split it in the ratio of 70% and 30%, where 70% is used to create the training dataset and the remaining 30% is used to create the test dataset. Then, we pass the training dataset to the ML system for training and use the test dataset to calculate the performance of the model. A train-test split is often performed in a random state, meaning 70% of the data that was used to create the training dataset is often chosen at random from the original dataset without replacement, except in the case of time-series data, where the order of the events needs to be maintained or in the case where we need to keep the classes stratified. Similarly, for the test dataset, 30% of the data is chosen at random from the original dataset to create the test dataset.

The following diagram shows how data from the dataset is randomly picked to create the training and testing datasets for their respective purposes:

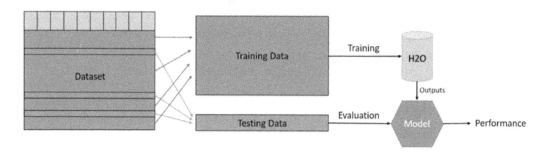

Figure 8.7 – Train-test split on the dataset

Now, the issue with the train-test split is that when 30% of the data kept outside the testing dataset is not used to train the model, any missing knowledge that could be derived from this data isn't available to train the model. This leads to a loss in the performance of the model. If you retrain a model using a different random state for the train-test split, then the model will end up having a different performance level as it has been trained on different data records. Thus, the performance of the models depends on the random assignment of the training dataset. So, how can we provide the test data for training as well as keeping some test data for performance measurement? This is where cross-validation comes into play.

Understanding cross-validation

Cross-validation is a model validation technique that resamples data to train and test models. The technique uses different parts of the dataset for training and testing during each iteration. Multiple iterations of model training and testing are performed using different parts of the dataset. The performance results are combined to give an average estimation of the model's performance.

Let's try to understand this with an example. Let's assume your dataset contains around 1,000 records. To perform cross-validation, you must split the dataset into a ratio – let's assume a 1:9 ratio where we have 100 records for the test dataset and 900 records for the training dataset. Then, you perform model training on the training dataset. Once the model has been trained, you must test the model on the test dataset and note its performance. This is your first iteration of cross-validation.

In the next iteration, you split the dataset in the same ratio of 1/9 records for the testing and training datasets, respectively, but this time, you choose different data records to form your test dataset and use the remaining records as the training dataset. Then, you perform model training on the training dataset and calculate the model's performance on the testing dataset. You repeat the same experiment using different data records until all the dataset has been used for training as well as testing. You will need to perform around 10 iterations of cross-validation so that, during the entire cross-validation process, the model is trained and tested on the entire dataset each iteration while containing different data records in the testing DataFrame.

Once all the iterations have finished, you must combine the performance results of the experiments and provide the average estimation of the model's performance. This technique is called cross-validation.

You may have noticed that during cross-validation, we perform model training multiple times on the same dataset. This is expected to increase the overall ML process time. This is especially true when performing cross-validation on a large dataset with a very high ratio between the training and testing partition. For example, if we have a dataset that contains 30,000 rows and we split the dataset into 29,000 rows for training and 1,000 rows for testing, then this will lead to a total of 3,000 iterations of model training and testing. Hence, there is an alternative form of cross-validation that lets you choose how many iterations to perform: called **K-fold cross-validation**.

In K-fold cross-validation, you decide the value of **K**, which is used to determine the number of cross-validation iterations to perform. Depending on the value of K, the ML service will randomly partition the dataset into K equal subsamples that will be resampled over the cross-validation iterations. The following diagram will help you understand this:

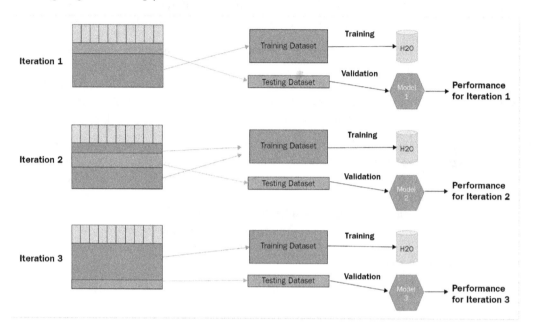

Figure 8.8 – K-fold cross-validation where K=3

As you can see, we have a dataset that contains 30,000 data records and the chosen value of K in the K-fold cross-validation is 3. Accordingly, the dataset will be split into 20,000 records for the test dataset and 10,000 records for training, which will be resampled during the following iterations, leading to a total of three cross-validations.

The benefit of using K-fold cross-validation to perform your model validation is that the model is trained on the entire dataset without missing out on data during training. This is especially beneficial in multi-class classification problems where there are chances that the model might miss out training on some of the prediction classes because it got split out from the training dataset to be used in the testing dataset.

Now that we have a better understanding of the basics of cross-validation and how it works, let's see how we can perform it using special parameters in the H2O AutoML training function.

Working with cross-validation parameters in H2O AutoML

H2O AutoML has provisions for you to implement K-fold cross-validation on your data for all ML algorithms that support it, along with some additional information that may help support the implementation.

You can use the following parameters to implement cross-validation:

- `nfolds`: This parameter sets the number of folds to use for K-fold cross-validation.

  In H2O Flow, this parameter will be available in the **ADVANCED** section of the **Run AutoML** parameters, as shown in the following screenshot:

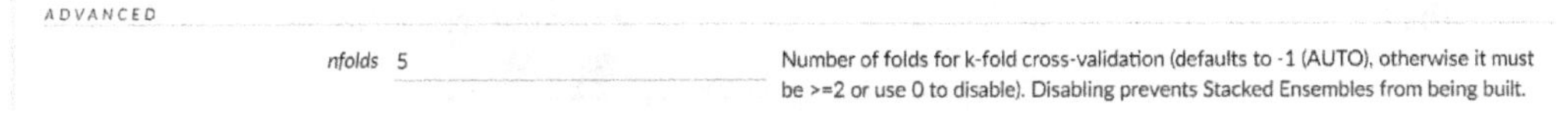

ADVANCED

nfolds 5   Number of folds for k-fold cross-validation (defaults to -1 (AUTO), otherwise it must
                                                               be >=2 or use 0 to disable). Disabling prevents Stacked Ensembles from being built.

Figure 8.9 – The nfolds parameter in H2O Flow

- `fold_assignment`: This parameter is used to specify the fold assignment scheme to use to perform K-fold cross-validation. The various types of fold assignments you can set are as follows:

  - `AUTO`: This assignment value lets the model training algorithm choose the fold assignment to use. `AUTO` currently uses `Random` as the fold assignment.

  - `Random`: This assignment value is used to randomly split the dataset based on the `nfolds` value. This value is set by default if `nfolds > 0` and `fold_column` is not specified.

  - `Modulo`: This assignment value is used to perform a modulo operation when splitting the folds based on the `nfolds` value.

  - `Stratified`: This assignment value is used to arrange the folds based on the response variable for classification problems.

  In the Python programming language, you can set these parameters as follows:

```
aml = h2o.automl.H2OAutoML(nfolds = 10, fold_assignment =
"AUTO", seed = 123)
aml.train(x = features, y = label, training_frame =
train_dataframe)
```

In the R programming language, you can set these parameters as follows:

```
aml <- h2o.automl(x = features, y = label, training_
frame = train_dataframe, seed = 123, nfolds = 10, fold_
assignment = "AUTO")
```

- **fold_column**: This parameter is used to specify the fold assignment based on the contents of a column rather than any procedural assignment technique. You can custom set the fold values per row in the dataset by creating a separate column containing the fold IDs and then setting **fold_column** to the custom column's name.

  In H2O Flow, this parameter will be available in the **ADVANCED** section of the **Run AutoML** parameters, as shown in the following screenshot:

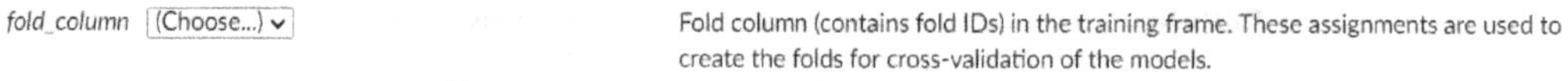

  fold_column (Choose...) ⌄   Fold column (contains fold IDs) in the training frame. These assignments are used to create the folds for cross-validation of the models.

  Figure 8.10 – The fold_column parameter in H2O Flow

  In the Python programming language, you can set these parameters as follows:

  ```
  aml.train(x = features, y = label, training_frame =
  train_dataframe, fold_column = "fold_column_name")
  ```

  In the R programming language, you can set these parameters as follows:

  ```
  aml <- h2o.automl(x = features, y = label, training_
  frame = train_dataframe, seed = 123, fold_column="fold_
  numbers")
  ```

- **keep_cross_validation_predictions**: When performing K-fold cross-validation, H2O will train *K+1* number of models, where *K* number of models are trained as a part of cross-validation and *1* additional model is trained on the entire dataset. Each of the cross-validation models makes predictions on the test DataFrame for that iteration and the predicted values are stored in a prediction frame. You can save these prediction frames by setting this parameter to *True*. By default, this parameter is set to *False*.

- **keep_cross_validation_models**: Similar to **keep_cross_validation_predictions**, you can also choose to keep the models trained during cross-validation for further inspection and experimentation by enabling this parameter to *True*. By default, this parameter is set to *False*.

- **keep_cross_validation_fold_assignment**: During cross-validation, the data is split either by the **fold_cloumn** or **fold_assignment** parameter. You can save the fold assignment that was used in cross-validation by setting this parameter to *True*. By default, this parameter is set to *False*.

In H2O Flow, these parameters will be available in the **EXPERT** section of the **Run AutoML** parameters, as shown in the following screenshot.

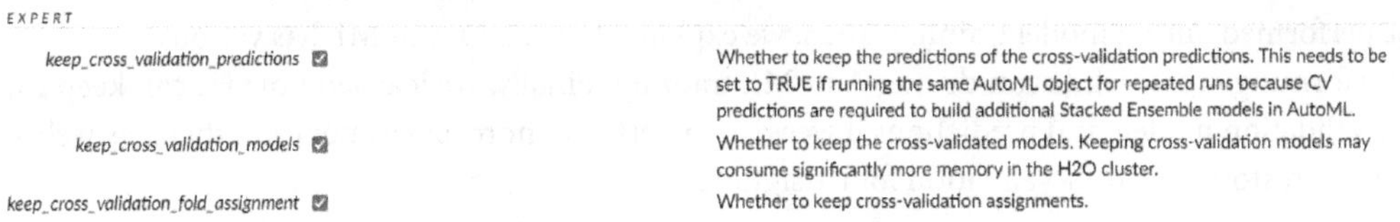

EXPERT

keep_cross_validation_predictions ☑	Whether to keep the predictions of the cross-validation predictions. This needs to be set to TRUE if running the same AutoML object for repeated runs because CV predictions are required to build additional Stacked Ensemble models in AutoML.
keep_cross_validation_models ☑	Whether to keep the cross-validated models. Keeping cross-validation models may consume significantly more memory in the H2O cluster.
keep_cross_validation_fold_assignment ☑	Whether to keep cross-validation assignments.

Figure 8.11 – Advanced cross-validation parameters in H2O Flow

In the Python programming language, you can set these parameters as follows:

```
aml = h2o.automl.H2OAutoML(nfolds = 10, keep_cross_
validation_fold_assignment = True, keep_cross_validation_
models = True, keep_cross_validation_predictions= True,
seed = 123)
aml.train(x = features, y = label, training_frame =
train_dataframe)
```

In the R programming language, you can set these parameters as follows:

```
aml <- h2o.automl(x = features, y = label, training_frame
= train_dataframe, seed = 123, nfolds = 10, keep_cross_
validation_fold_assignment = TRUE, keep_cross_validation_
models = TRUE, keep_cross_validation_predictions= TRUE)
```

Congratulations – you have now understood a few more advanced ML concepts and how to use them in H2O AutoML!

Summary

In this chapter, we learned about some of the optional parameters that are available to us in H2O AutoML. We started by understanding what imbalanced classes in a dataset are and how they can cause trouble when training models. Then, we understood oversampling and undersampling, which we can use to tackle this. After that, we learned how H2O AutoML provides parameters for us to control the sampling techniques so that we can handle imbalanced classes in datasets.

After that, we understood another concept, called early stopping. We understood how overtraining can lead to an overfitted ML model that performs very poorly against unseen new data. We also learned that early stopping is a method that we can use to stop model training once we start noticing that the model has started overfitting by monitoring the performance of the model against the validation dataset. We then learned about the various parameters that H2O AutoML has that we can use to automatically stop model training once overfitting occurs during model training.

Next, we understood what cross-validation is and how it helps us train the model on the entire dataset, as well as validate the model's performance as if the model had seen the data for the first time. We also learned how K-fold cross-validation helps us control the number of cross-validation iterations to be performed during model training. Then, we explored how H2O AutoML has various provisions for performing cross-validation during AutoML training. Finally, we learned how we can keep the cross-validation models and predictions if we wish to perform more experiments on them, as well as how we can store the cross-validation fold assignments.

In the next chapter, we shall explore some of the miscellaneous features that H2O AutoML has that can be useful to us in certain scenarios.

9

Exploring Miscellaneous Features in H2O AutoML

Along with incorporating many **Machine Learning** (**ML**) algorithms and various features to train them, H2O AutoML has a few miscellaneous features that make it an all-around service capable of catering to all kinds of business requirements.

H2O AutoML's strength not only lies in its ability to train multiple models automatically but also in providing support for other services and features that are vital for production-grade systems.

In this chapter, we shall explore two unique features of H2O AutoML that are good to know and that can be very useful when required. The first one is H2O AutoML's compatibility with a popular ML library in Python called scikit-learn. We shall explore how we can use H2O AutoML in a scikit-learn implementation and how it can provide value to the large scikit-learn community.

The second feature is an inbuilt logging system in H2O AutoML. This logging system logs valuable information during the AutoML training process. It can be especially useful if you plan to use the H2O AutoML service in production, where monitoring the health of your systems is of utmost priority.

In this chapter, we will cover the following topics:

- Understanding H2O AutoML integration in scikit-learn
- Understanding H2O AutoML event logging

With this in mind, let's explore H2O AutoML's first miscellaneous feature compatibility with scikit-learn.

Technical requirements

For this chapter, you will require the following:

- The latest version of your preferred web browser

- An **Integrated Development Environment (IDE)** of your choice

- (Optional) Jupyter Notebook by Project Jupyter (`https://jupyter.org/`)

All the experiments conducted in this chapter have been performed on Jupyter notebooks to provide you with better visual examples of outputs. You are free to follow along using the same setup. You can also perform the same experiments on any Python environment as the Python code will execute the same on both environments.

All the code examples for this chapter can be found on GitHub at `https://github.com/PacktPublishing/Practical-Automated-Machine-Learning-on-H2O/tree/main/Chapter%209`.

Understanding H2O AutoML integration in scikit-learn

Scikit-learn is one of the most commonly used open source ML libraries in the field of ML and data science. It is a library for the Python programming language and focuses on ML tooling functions. It involves modules that perform mathematical and statistical analysis, general-purpose ML algorithms, as well as functions to train, test, and evaluate ML models.

Scikit-learn was originally developed by David Cournapeau and was initially called **scikits.learn**. It was created as a Google Summer of Code project in 2007, which was later picked up as a thesis project by Matthieu Brucher that same year. It was later re-written and further developed by Fabian Pedregosa, Gael Varoquaux, Alexandre Gramfort, and Vincent Michel from the French Institute of Research in Computer Science and Automation in Rocquencourt, France. Scikit-learn's first public release of version 1 was made on February 1, 2010.

You can find more details about scikit-learn here: `https://scikit-learn.org/stable/`.

The scikit-learn library is built on the following packages:

- **NumPy**: NumPy is a Python library used to work exclusively with arrays. It is used for scientific computing in Python and provides functions to work with multi-dimensional arrays. It also provides a wide variety of fast computing mathematical operations on said arrays, making it ideal for data analytics. It can perform array shape manipulation, sorting, searching, discrete Fourier transform operations, linear algebra, and statistics. You can find more details about NumPy here: `https://numpy.org/`.

- **SciPy**: SciPy is a scientific computational library that is built on top of NumPy. It provides advanced scientific computational functions. It is used to perform operations such as image processing, clustering, gradient optimization, and much more. All numerical computations in Python are done by SciPy. You can find more details about SciPy here: `https://scipy.org/`.

- **Matplotlib**: Matplotlib is a library that is used for creating visualizations from data. These visualizations involve various types of graphs and charts that rely on computed data that can be easily expressed and explained visually. It can create plot diagrams that are good for publications in scientific research papers and create interactive diagrams, all of which can be exported into different types of formats. You can find more details about Matplotlib here: `https://matplotlib.org/`.

Scikit-learn is often used by data scientists when experimenting with data. It provides tons of flexibility when conducting experiments and since its APIs are very easy to use, it is often the go-to library for performing general ML functions.

H2O AutoML can be easily integrated with scikit-learn. You can use H2O AutoML as a scikit-learn **Estimator** and use it in conjunction with other scikit-learn functions to use the best of both worlds. H2O AutoML interacts with scikit-learn using the **h2o.sklearn** module. The h2o.sklearn module exposes two wrapper functions to perform AutoML:

- **H2OAutoMLClassifier**: This function is used to train classification models using H2O AutoML
- **H2OAutoMLRegressor**: This function is used to train regression models using H2O AutoML

The functions accept input data of various formats such as H2Oframes, NumPy arrays, or even pandas DataFrames. They also expose standard training and prediction APIs that are similar to how they are used in scikit-learn. This enables scikit-learn to use H2O AutoML, along with other scikit-learn components.

The H2O AutoML Estimators also retain their original functionality, such as leaderboards and training information, among others. Users can still access these details in scikit-learn to extract information from the AutoML training for further experimentation or analysis.

Now that we have a better understanding of the scikit-learn library and what it is used for, let's learn how to use it alongside H2O AutoML. We will start by understanding the various ways that we can install scikit-learn on our system.

Building and installing scikit-learn

Installing scikit-learn is very easy. There are three different ways to install scikit-learn on your system:

- Installing the latest official release of scikit-learn
- Installing the scikit-learn version provided by your Python distribution or operating system
- Building and installing the scikit-learn package from the source

Let's quickly go through these options one by one so that we have scikit-learn ready to use alongside H2O AutoML.

Installing the latest official release of scikit-learn

This process can vary, depending on what type of Python package manager you are using on your system:

- Using the `pip` package manager, execute the following command in your Terminal to install the latest release of scikit-learn:

```
pip install -U scikit-learn
```

  - The following command will show you where scikit-learn is installed, as well as its version:

```
python -m pip show scikit-learn
```

- Using the **Anaconda** or **Miniconda** package manager, execute the following command in your Terminal to install the latest release of scikit-learn:

```
conda create -n sklearn-env -c conda-forge scikit-learn
conda activate sklearn-env
```

The following command will show you the version of scikit-learn installed on your system:

```
conda list scikit-learn
```

You can use the following command to import the installed scikit-learn module to ensure that it is successfully installed and then display its version:

```
python -c "import sklearn; sklearn.show_versions()"
```

Now that we know how to install scikit-learn using `pip`, Anaconda, and Miniconda, let's look at another way of installing it using the Python distribution that comes packaged with your operating system.

Installing scikit-learn using your operating system's Python distribution

Since scikit-learn is so commonly used by developers, it is often packaged along with the built-in package manager in various Python distributions or operating systems. This enables users to directly install the available scikit-learn package without needing to download it from the internet.

The following is a list of some of the operating systems that come with their own version of prepackaged scikit-learn and the respective Terminal commands to install it:

- **Arch Linux**: The Arch Linux operating system distribution provides the scikit-learn library out of the box in the form of `python-scikit-learn`. To install this library, execute the following command:

```
sudo pacman -S python-scikit-learn
```

- **Debian/Ubuntu**: The Debian Ubuntu distribution splits the scikit-learn package into three parts:

  - **python3-sklearn**: This package contains the Python modules for scikit-learn functions
  - **python3-sklearn-lib**: This package contains the low-level implementations and bindings for scikit-learn
  - **python3-sklearn-doc**: This package contains the documentation for scikit-learn

  To install this library, execute the following command:

  ```
  sudo apt-get install python3-sklearn python3-sklearn-lib
  python3-sklearn-doc
  ```

- **Fedora**: The Fedora operating system distribution provides the scikit-learn library out of the box in the form of `python3-scikit-learn`. It is the only one available in `Fedora30`:

  ```
  sudo dnf install python3-scikit-learn
  ```

- **NetBSD**: In NetBSD, the scikit-learn library can be installed via its portable packaging system, called **pkgsrc-wip**. You can download the scikit-learn package for `pkgsrc-wip` from here: `http://pkgsrc.se/math/py-scikit-learn`.

The drawback of this process is that it often comes with an older version of scikit-learn. This, however, can be fixed by upgrading the installed packages to the latest versions using the respective package managers.

Building and installing the scikit-learn package from the source

Users that want to use the latest experimental features or those who wish to contribute to scikit-learn can directly build and install scikit-learn's latest available version.

You can build and install scikit-learn from the source by executing the following steps:

1. Use Git to check out the latest source from the scikit-learn repository on GitHub. The scikit-learn repository can be found here: `https://github.com/scikit-learn/scikit-learn`. Execute the following command to clone the latest scikit-learn repository:

   ```
   git clone git://github.com/scikit-learn/scikit-learn.git
   ```

2. Use Python to create a virtual environment and install **NumPy**, **SciPy**, and **Cython**, which are the build dependencies for scikit-learn:

   ```
   python3 -m venv h2o-sklearn
   source h2o-sklearn/bin/activate
   pip install wheel numpy scipy cython
   ```

3. Use `pip` to build the project by running the following command:

```
pip install --verbose --no-build-isolation --editable .
```

4. Once the installation is completed, check if scikit-learn is installed correctly by running the following command:

```
python -c "import sklearn; sklearn.show_versions()"
```

To avoid conflicts with other packages, it is highly recommended to install scikit-learn in a virtual environment or a **conda** environment. Also, when installing SciPy and NumPy, it is recommended to use **binary wheels** as they are not recompiled from the source.

Experimenting with scikit-learn

Now that we have successfully installed scikit-learn, let's quickly look at a simple implementation of scikit-learn for training a model. Using this as a reference, we shall then explore how we can incorporate H2O AutoML into it.

The dataset we shall use for this experiment will be the same Iris flower dataset that we have been using throughout this book. This dataset is a good example of using ML to solve a classification problem.

So, let's begin by implementing it using pure scikit-learn functions.

Follow these steps to train your ML model in Python using scikit-learn:

1. Import the `sklearn` and `numpy` libraries:

```
import sklearn
import numpy
```

2. The Iris flower dataset is readily available in the `sklearn` library; it is present in the dataset submodule of `sklearn`. Next, import that dataset by executing the following commands. Let's also have a closer look at the contents of the DataFrame:

```
from sklearn.datasets import load_iris
dataframe = load_iris()
print(dataframe)
```

You should get an output displaying the contents of the DataFrame in the form of a dictionary. Let's investigate the important key-value pairs in the dictionary to understand what we are dealing with:

-   `data`: This key contains all the features of the dataset – that is, the sepal length, sepal width, petal length, and petal width – in the form of a multi-dimensional array.

- target_names: This key contains the names of the target or labels of the dataset – that is, Iris-setosa, Iris-versicolour, and Iris-virginica. This is an array, and the index of the names is the numerical representation that is used in the actual content of the dataset.

- target: This key contains all the target values, also called label values, of the dataset. This is also an array that represents the values of the target that would have otherwise been a column in a tabular dataset. The values are numeric, where 0 represents Iris-setosa, 1 represents Iris-versicolour, and 2 represents Iris-virginica, as decided by their index values in target_names.

3. With this information in mind, extract the features and labels into separate variables by executing the following command:

```
features = dataframe.data
label = dataframe.target
```

4. We need to split the dataset into two parts – one for training and the other for testing. Unlike H2O, in scikit-learn, we treat the features and labels as two separate entities. Both of them should have the same dimensional length to match the data contents. To do this split, execute the following commands:

```
from sklearn.model_selection import train_test_split
feature_train, feature_test, label_train, label_test =
train_test_split(features, label, test_size=0.30, random_
state=5)
```

The split functionality splits the features and labels into a 70% to 30% ratio, where 70% of the data is kept for training and the remaining 30% is kept for testing. So, we eventually end up with a total of four DataFrames, as follows:

- feature_train: This DataFrame contains 70% of the feature data to be used for training

- label_train: This DataFrame contains 70% of the label data to be used for training

- feature_test: This DataFrame contains 30% of the feature data to be used for testing

- label_test: This DataFrame contains 30% of the label data to be used for testing

5. Once the training and testing DataFrames are ready, declare and initialize the ML algorithm to be used for model training. Scikit-learn has separate libraries for different types of algorithms. Since we are working with a classification problem, let's use the **logistic regression** algorithm to train a classification model. Execute the following command to initialize a logistic regression function to train a model:

```
from sklearn.linear_model import LogisticRegression
logReg = LogisticRegression(solver='lbfgs', max_
iter=1000)
```

6. Now, let's train a model using the `feature_train` and `label_train` datasets. Execute the following function:

```
logReg.fit(feature_train, label_train)
```

7. Once training is finished, we can use the same logistic regression object to make predictions on the `feature_test` DataFrame. Execute the following command and print out the prediction's output:

```
predictions = logReg.predict(feature_test)
print(predictions)
```

You should get an output similar to the following:

```
[1 2 2 0 2 1 0 2 0 1 1 2 2 2 0 0 2 2 0 0 1 2 0 1 1 2 1 1 1 2 0 1 1 0 1 0 0
 2 0 2 2 1 0 0 1]
```

Figure 9.1 – Prediction output from scikit-learn logistic regression

8. You can also measure the accuracy of your predictions by executing the following command:

```
score = logReg.score(feature_test, label_test)
print(score)
```

You should get an accuracy of around 97.77.

In this experiment, we learned how to use scikit-learn to import a dataset, perform splitting, and then use logistic regression to train a classification model. But as we learned in the previous chapters, there are plenty of ML algorithms to choose from. Each has its own way of dealing with **variance** and **bias**. So, as expected, the most obvious question remains unanswered: *which ML algorithm should we use?*

As we saw in this experiment, scikit-learn may have tons of support for different algorithms, but training all of them can become complicated from a programming point of view. This is where we can integrate H2O AutoML to perform automated model training to train all the ML algorithms.

Now that we have got a good idea of how we can use scikit-learn to train models, let's see how we can use H2O AutoML with scikit-learn.

Using H2O AutoML in scikit-learn

First, we will learn how to use H2O AutoML in scikit-learn to perform classification using the `H2OAutoMLClassifier` submodule. We shall use the same classification ML problem using the Iris dataset and see how we can train multiple models using H2O AutoML.

Experimenting with H2OAutoMLClassifier

Follow these steps to train your H2O AutoML classification model in Python using scikit-learn:

1. Implement *steps 1* to *4* that we followed in the *Experimenting with scikit-learn* section.

2. In the experiment we performed in the *Experimenting with scikit-learn* section, after *step 4*, we initialized the logistic regression algorithm by importing the LogisticRegression submodule from sklearn.linear_model. In this experiment, we will import the H2OAutoMLClassifier submodule from the h2o.sklearn module instead:

    ```
    from h2o.sklearn import H2OAutoMLClassifier
    h2o_aml_classifier = H2OAutoMLClassifier(max_
    models=10, seed=5, max_runtime_secs_per_model=30, sort_
    metric='logloss')
    ```

    Just like how we set the AutoML parameters in the previous chapters, we have set max_models to 10, max_runtime_secs_per_model to 30 seconds, the random seed value to 5, and sort_metric to logloss.

3. Once H2OAutoMLClassifier has been initialized, you can use it to fit – in other words, train – your models. Execute the following command to trigger AutoML training:

    ```
    h2o_aml_classifier.fit(feature_train, label_train)
    ```

    First, the program will check if an H2O instance is already running on localhost:54321. If not, then H2O will spin up an instance of the H2O server; otherwise, it will reuse the already existing one to train the AutoML models. Once training starts, you should get an output similar to the following:

Figure 9.2 – Output from H2O AutoML classifier training

From the output, you can see that H2O first imported and parsed the **feature_train** and **label_train** DataFrames. Then, it started the AutoML training.

4. To view the results of the AutoML training, you can view the H2O **leaderboard** by executing the following command:

```
h2o_aml_classifier.estimator.leaderboard
```

You should get an output similar to the following:

model_id	logloss	mean_per_class_error	rmse	mse
GLM_1_AutoML_2_20220613_224223	0.0860788	0.0479303	0.162001	0.0262443
DeepLearning_grid_1_AutoML_2_20220613_224223_model_1	0.112033	0.038671	0.170577	0.0290966
StackedEnsemble_AllModels_1_AutoML_2_20220613_224223	0.11419	0.0378462	0.176666	0.0312109
StackedEnsemble_BestOfFamily_1_AutoML_2_20220613_224223	0.120517	0.0476502	0.18068	0.0326452
XRT_1_AutoML_2_20220613_224223	0.124044	0.0473856	0.190803	0.0364056
GBM_2_AutoML_2_20220613_224223	0.132214	0.0473856	0.200067	0.040027
DRF_1_AutoML_2_20220613_224223	0.133123	0.0473856	0.189798	0.0360231
GBM_3_AutoML_2_20220613_224223	0.147124	0.0473856	0.202676	0.0410774
GBM_4_AutoML_2_20220613_224223	0.155193	0.0473856	0.20303	0.041221
GBM_5_AutoML_2_20220613_224223	0.184402	0.0473856	0.210164	0.0441689

Figure 9.3 – H2O AutoML leaderboard

5. Using the same H2O AutoML classifier, you can also make predictions, as follows:

```
predictions = h2o_aml_classifier.predict(feature_test)
print(predictions)
```

You should get an output similar to the following:

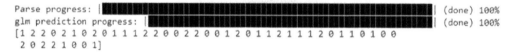

Figure 9.4 – Output of the prediction using H2OAutoMLClassifier

By default, the classifier will use the model with the highest rank on the leaderboard to make predictions.

With that, you have learned how to implement H2O AutoML in scikit-learn to solve classification problems using H2OAutoMLClassifier.

Now that we have a good idea of how we can use H2OAutoMLClassifier to perform classification predictions on data, let's see how we can perform regression predictions using the H2OAutoMLRegressor submodule.

Experimenting with H2OAutoMLRegressor

Now, let's see how we solve a **regression** problem using H2OAutoMLRegressor. For this experiment, we shall use the Red Wine Quality dataset that we used previously in *Chapter 7, Working with Model Explainability*.

Follow these steps to train your H2O AutoML regression model in Python using scikit-learn:

1. Implement *Steps 1* to *4* that we followed in the *Experimenting with scikit-learn* section.

2. In the experiment we performed in the *Experimenting with H2OAutoMLClassifier* section, we initialized H2OAutoMLClassifier. Since we are dealing with a regression problem in this experiment, we shall use the H2OAutoMLRegressor submodule. Execute the following command to import and instantiate the H2OAutoMLRegressor class object:

    ```
    from h2o.sklearn import H2OAutoMLRegressor
    h2o_aml_regressor = H2OAutoMLRegressor(max_models=10,
    max_runtime_secs_per_model=30, seed=5)
    ```

3. Once H2OAutoMLRegressor has been initialized, we can trigger AutoML to train our regression models. Execute the following command to trigger AutoML:

    ```
    h2o_aml_regressor.fit(feature_train, label_train)
    ```

    Once model training finishes, you should get an output similar to the following:

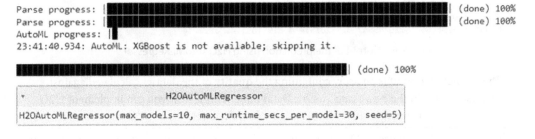

Figure 9.5 – Output from H2O AutoML regressor training

4. Similar to H2OAutoMLClassifier, you can also view the results of the AutoML training on the H2O **leaderboard** by executing the following command:

    ```
    h2o_aml_regressor.estimator.leaderboard
    ```

5. Making predictions is also very easy. You use the same H2OAutoMLRegressor object and call its predict method while passing the feature dataset kept aside for testing. Execute the following command to make predictions using the leader model trained by H2OAutoMLRegressor:

```
predictions = h2o_aml_regressor.predict(feature_test)
print(predictions)
```

You should get the results of the prediction, as follows:

```
Parse progress: |████████████████████████████████| (done) 100%
stackedensemble prediction progress: |████████████████████████| (done) 100%
[5.7050527  4.71579041 4.99625922 6.88799804 5.30901742 5.08507949
 5.22402198 6.80312639 5.73784549 5.05679221 6.60598185 5.16579878
 5.14112093 5.8436239  6.03287821 5.95839962 6.33995783 6.05639611
 5.76648316 5.75068268 4.90408778 5.29082171 5.17198907 5.217705
 5.97337232 5.43873863 5.19827529 5.31316769 6.2793359  6.38171303
 5.58378112 5.36376057 5.42070109 4.84603301 5.9211492  6.17345933
 6.09528051 5.68300139 5.10375214 5.33199332 6.11587125 5.05826719
 5.52626887 5.12766708 6.54879736 5.67341307 5.24367939 5.7050527
 5.72492442 5.20803101 5.72238569 5.23847352 4.94162759 5.72515368
 5.99710415 5.54065627 5.96809731 5.41532481 6.51548553 6.43069508
 5.41738151 5.50117357 6.77200607 5.60270773 5.02891041 6.637265
 6.61085364 5.51616503 5.77759467 5.88147359 5.12760658 6.87250045
 5.61076927 5.45198623 5.4550793  5.02793331 4.86960596 5.24755881
 5.14942367 5.03725484 5.36312455 4.98812026 5.49151424 5.16758049
 5.04899969 5.48321742 5.63502592 5.01813754 5.63146962 5.06783828
 6.08415337 6.10573536 5.89912238 5.51682489 5.8902318  5.61761149
 5.83461759 5.45310272 5.81230101 6.56012283 5.68300139 5.72515368
 5.42014689 6.37008337 5.44917672 5.65400694 5.37099615 5.05670911
 5.50945522 4.7857324  6.07391301 5.6309158  5.38736267 5.94324949
 6.12127789 5.25270546 5.09663222 4.94148322 5.24268557 5.76740135
 5.02992971 4.98601981 5.23574556 6.31680286 4.96510575 5.52742118
 5.05826719 5.69985666 5.16785547 6.83281514 4.81951631 4.98821234
```

Figure 9.6 – Output of the prediction using H2OAutoMLRegressor

The prediction output is an array containing the **quality** of the dataset calculated against each data entry in the feature_test DataFrame. This is how you can implement H2O AutoML in scikit-learn to solve regression problems using H2OAutoMLRegressor.

Now that you know how to use H2O AutoML in scikit-learn, let's move on to the next miscellaneous feature of H2O AutoML: event logging.

Understanding H2O AutoML event logging

Since H2O AutoML automates most of the ML process, we have given some control to the machine. Encapsulation means that all the complexities that lie in AutoML are all hidden away, and we are just aware of the inputs and whatever output H2O AutoML gives us. If there is any issue in H2O AutoML and it gives us models that don't make sense or are not expected, then we will need to dig deeper into how AutoML trained the models. Hence, we need a way to keep track of what's happening internally in H2O AutoML and whether it is training models as expected or not.

When building such software systems that are aimed to be used in production, you will always need a logging system to log information. The virtual nature of software makes it difficult for users to keep track of what is going on as the system does its processing and other activities. Any failures or issues can lead to a cascade of underlying problems that developers may end up finding out too late, if ever.

That is why logging systems are always implemented to provide support to your system. Logs generated by your system help developers track down the source of the problem and mitigate it promptly. H2O AutoML can also generate logs containing meta-information about all the underlying processing that happens when it is training models. You can use these logs to keep some sense of control when you are letting H2O take care of all the ML processing.

There are two types of logs that AutoML generates. They are as follows:

- **Event Logs**: These are event logs that are generated in the backend of AutoML as training progresses. All the logs are collected and presented as an H2O DataFrame.

- **Training Logs**: These are logs that show training and prediction times as AutoML trains models and are in the form of a dictionary of key-value pairs. The training times are in epochs and are mostly useful for post-analysis of model training.

Let's see how we can retrieve these logs from H2O AutoML via a practical implementation.

Follow these steps to train models using H2O AutoML. Then, we will learn how to extract the logs and understand what they look like:

1. Import the h2o module and initialize H2O to spin up a local H2O server:

```
import h2o
h2o.init()
```

2. Import the Iris dataset by passing the location of where you downloaded the dataset:

```
data = h2o.import_file("Dataset/iris.data")
```

3. Set the label and features:

```
label = "C5"
features = data.columns
features.remove(label)
```

4. Initialize the H2O AutoML object with parameters, as follows:

```
aml = h2o.automl.H2OAutoML(max_models=10, seed = 5)
```

5. Trigger AutoML training by passing in the feature columns, label column, and the DataFrame to use to train the models on:

```
aml.train(x = features, y = label, training_frame =
dataframe)
```

6. Once training is finished, you can view the event logs by using the `event_log` property of the AutoML object. Let's retrieve the log DataFrame and have a look at its content:

```
event_logs = aml.event_log
print(event_logs)
```

You should get an output similar to the following:

timestamp	level	stage	message	name	value
00:13:05.18	INFO	Workflow	Project: AutoML_2_20220615_01305		
00:13:05.19	INFO	Validation	5-fold cross-validation will be used.		
00:13:05.19	INFO	Validation	Setting stopping tolerance adaptively based on the training frame: 0.05		
00:13:05.19	INFO	Validation	Build control seed: 5		
00:13:05.19	INFO	DataImport	training frame: Frame key: AutoML_2_20220615_01305_training_iris_data1.hex cols: 5 rows: 150 chunks: 1 size: 1982 checksum: -5547756281708519194		
00:13:05.19	INFO	DataImport	validation frame: NULL		
00:13:05.19	INFO	DataImport	leaderboard frame: NULL		
00:13:05.19	INFO	DataImport	blending frame: NULL		
00:13:05.19	INFO	DataImport	response column: C5		
00:13:05.19	INFO	DataImport	fold column: null		

Figure 9.7 – Event log output from H2O AutoML

Similarly, you can view the event logs in the R programming language by executing the following R commands:

```
event_log <- aml@event_log
```

The event log contains the following information:

- **timestamp**: This column is the time of occurrence of a particular event.

- **level**: In logging systems, logs are generally categorized into certain classes of importance or criticality. In most cases, the levels are as follows based on criticality ranking:

    i.   **FATAL**: This log level indicates that the application is facing a critical issue and will need to stop functioning and shut down.

    ii. **ERROR**: This log level indicates that the application is facing an issue performing certain functions. However, the issue is not that critical that the application needs to shut down.

    iii. **WARN**: This log level indicates that the application has detected something unusual that is harmless and is not affecting any functionality.

    iv. **INFO**: This log level indicates normal behavior updates that can be recorded and stored for future reference if needed. They are usually informative.

v. **DEBUG**: This log level indicates more detailed diagnostic details that are often needed when developing an application or to gather more information when you are performing diagnostic actions or debugging an issue.

vi. **TRACE**: This log level is similar to **DEBUG** albeit with finer details, especially when you are tracing the flow of information in a code base.

- **stage**: This column indicates the stage in AutoML training at which the log was generated.

- **message**: This column contains a descriptive message that provides information about the event that occurred.

- **name**: This column contains the name of the event log that occurred if it is set.

- **value**: This column contains the value of the event log that occurred if it is set.

7. Now, let's retrieve the training logs and look at their content:

```
info_logs = aml.training_info
print(info_logs)
```

You should get an output similar to the following:

```
{'creation_epoch': '1655248385',
 'start_epoch': '1655248385',
 'start_GLM_def_1': '1655248385',
 'start_GBM_def_5': '1655248386',
 'start_DRF_def_1': '1655248386',
 'start_GBM_def_2': '1655248386',
 'start_GBM_def_3': '1655248386',
 'start_GBM_def_4': '1655248386',
 'start_DRF_XRT': '1655248386',
 'start_GBM_def_1': '1655248387',
 'start_DeepLearning_def_1': '1655248387',
 'start_GBM_grid_1': '1655248387',
 'start_DeepLearning_grid_1': '1655248387',
 'start_StackedEnsemble_best_of_family_xglm': '1655248395',
 'start_StackedEnsemble_all_xglm': '1655248396',
 'stop_epoch': '1655248397',
 'duration_secs': '12'}
```

Figure 9.8 – Event log output from H2O AutoML

Similarly, you can view the event logs in the R programming language by executing the following R commands:

```
info_logs <- aml@training_info
```

The training log contains the following information:

- `creation_epoch`: This key in the training log dictionary contains the epoch value of when the AutoML job was created.

- `start_epoch`: This key in the training log dictionary contains the epoch value of when the AutoML build started.

- `start_{model_name}`: This type of key in the training log dictionary contains the epoch value of when the training of a particular model was started.

- `stop_epoch`: This key in the training log dictionary contains the epoch value of when the AutoML build stopped.

- `duration_secs`: This key in the training log dictionary contains the total time when AutoML was running in seconds.

This experiment gives us a good example of how H2O generates log events. When building an ML system using H2O AutoML, you can incorporate these logs into your logging system to keep an eye on H2O AutoML's functionality. This will help you identify any issues that may arise promptly and keep your models in production of the highest quality. You will be alerted if there were any issues during training and before you accidentally end up deploying faulty models in production.

Summary

In this chapter, we understood some of the miscellaneous features of H2O AutoML. We started by understanding the scikit-learn library and getting an idea of its implementation. Then, we saw how we can use the `H2OAutoMLClassifier` library and the `H2OAutoMLRegressor` library in a scikit-learn implementation to train AutoML models.

Then, we explored H2O AutoML's logging system. After that, we implemented a simple experiment where we triggered AutoML training; once it was finished, we extracted the event logs and the training logs in both the Python and R programming languages. Then, we understood the contents of those logs and how that information benefits us in keeping an eye on H2O AutoML functionality.

In the next chapter, we shall further focus on using H2O in production and how we can do so using H2O's Model Object Optimized.

10
Working with Plain Old Java Objects (POJOs)

Companies often use a mix of strategies that can deliver services up to the expected standards. In the case of services that use **Machine Learning** (**ML**), they need to consider how they can quickly and easily build, extract, and deploy their models in production without affecting their ongoing service.

Hence, the portability of trained models is very important. How do you take a model object created by your training pipeline built with a certain technology and use that in your prediction pipeline, which might be built using a different technology? Ideally, the model object should be an object that is self-contained and easily distributable.

In the world of software engineering, the Java programming language has been known to be one of the most widely used platform-independent programming languages. When Java compiles a program, it converts it into platform-independent byte code that can be interpreted by any machine that has a **Java Virtual Machine** (**JVM**) installed in it. And expanding on this feature, you have **Plain Old Java Objects** (**POJOs**).

POJOs are ordinary objects that can be run by any Java program, irrespective of any framework. This makes POJOs very portable when deployed to different kinds of machines. H2O also has provisions to extract trained models in the form of POJOs, which can then be used for deployment in production.

In this chapter, we shall dive deep into understanding what POJOs are and how we can download them after successfully training a model in Python, R, and H2O Flow. Then, we'll learn how to load a POJO into a simple Java program to make predictions.

In this chapter, we will cover the following topics:

- Introduction to POJOs
- Extracting H2O models as POJOs
- Using a H2O model as a POJO

By the end of this chapter, you should be able to extract trained models in the form of POJOs using Python, R, or H2O Flow and then load these POJO models into your ML program to make predictions.

Technical requirements

For this chapter, you will require the following:

- The latest version of your preferred web browser.

- An **Integrated Development Environment (IDE)** of your choice.

- (Optional) Jupyter Notebook by Project Jupyter (`https://jupyter.org/`)

  All the experiments conducted in this chapter are performed on Jupyter notebooks to provide you with better visual examples of outputs. You are free to follow along using the same setup or perform the same experiments in environments specific to the language you are using. All the code examples for this chapter can be found on GitHub at `https://github.com/PacktPublishing/Practical-Automated-Machine-Learning-on-H2O/tree/main/Chapter%2010`.

Introduction to POJOs

POJO is a term coined by Martin Fowler, Rebecca Parsons, and Josh Mackenzie in September 2000. It is an ordinary Java object, but what makes it *plain old* is not what it should do but rather what it should not do.

A Java object can be a POJO in the following circumstances:

- The Java object does not extend from any class.

- The Java object does not implement any interfaces.

- The Java object does not use any annotations from outside.

What these three restrictions lead to is a Java object that is not dependent on any other library or object outside of itself and is self-contained d enough to perform its logic on its own. You can easily embed POJOs in any Java environment due to their portability, and because of Java's platform independence, they can be run on any machine.

H2O can export trained models in the form of POJOs. These POJO models can then be deployed and used to make predictions on inbound data. The only dependency on using POJO models is the `h2o-genmodel.jar` file. This is a JAR file that is needed to compile and run H2O model POJOs. This JAR file is a library that contains the base classes and `GenModel`, a helper class to support Java-generated models, from which the model POJOs are derived. This same library is also responsible for supporting scoring by using the model POJOs.

When working with model POJOs in production, you will need the `h2o-genmodel.jar` file to compile, deploy, and run your model POJOs. POJOs are simple Java code that are not tied to any particular version of H2O. However, it is still recommended to use the latest version of `h2o-genmodel.jar` since it can load the current version, as well as older versions, of your POJO. You can find detailed documentation regarding `h2o-genmodel.jar` at `https://docs.h2o.ai/h2o/latest-stable/h2o-genmodel/javadoc/index.html`.

Now that we know what POJOs are and how H2O model POJOs work, let's learn how to extract trained H2O models using AutoML as POJOs by using simple examples.

Extracting H2O models as POJOs

Models trained using H2O's AutoML can also be extracted as POJOs so that they can be deployed to your production systems.

In the following sub-sections, we shall learn how to extract the model POJOs using the Python and R programming languages, as well as how we can extract model POJOs using H2O Flow.

Downloading H2O models as POJOs in Python

Let's see how we can extract H2O models as POJOs using a simple example in Python. We shall use the same Iris flower dataset we have been using so far. This dataset can be found at `https://archive.ics.uci.edu/ml/datasets/iris`.

Follow these steps to train models using H2O AutoML in Python. After doing this, you will extract the leader model and download it as a POJO:

1. Import the h2o module and start your H2O server:

   ```
   import h2o
   h2o.init()
   ```

2. Import the dataset by passing the location of the dataset in your system. Execute the following command:

   ```
   data_frame = h2o.import_file("Dataset/iris.data")
   ```

3. Set the feature and label names by executing the following commands:

   ```
   features = data_frame.columns
   label = "C5"
   features.remove(label)
   ```

4. Initialize the H2O AutoML object and set the `max_model` parameter to `10` and the `seed` value to 5 by executing the following commands:

    ```
    aml=h2o.automl.H2OAutoML(max_models=10, seed = 5)
    ```

5. Trigger AutoML by passing the training dataset, the feature columns, and the label column as the parameters, as follows:

    ```
    aml.train(x = features, y = label, training_frame = data_
    frame)
    ```

6. Once the training has finished, H2O AutoML should have trained a few models and ranked them based on a default ranking performance metric on a leaderboard. The highest ranking model on the leaderboard is called a *leader* and can be accessed directly by using the `aml.leader` command. Using this reference, you can download the leader model as a POJO by running the following command:

    ```
    h2o.download_pojo(aml.leader, path="~/Downloads/", jar_
    name="AutoMLModel")
    ```

    This should download a model POJO called `AutoMLModel`, as specified in the `jar_name` parameter, to the path specified in the `path` parameter. If the `path` parameter is not set, then H2O will print the model POJO's details on the console instead of downloading it as a JAR file.

You can also view the contents of the POJO by opening the file in any editor. The file will contain a single public class that is named after your leader model and extends the `GenModel` class, which is a part of `h2o-genmodel.jar`.

Now that we know how we can extract a POJO model using Python, let's see a similar example in the R programming language.

Downloading H2O models as POJOs in R

Similar to how we can extract a model from the AutoML leaderboard in Python, we can do the same in the R programming language. We shall use the same Iris flower dataset in this section. Follow these steps to train models using H2O AutoML and then extract the leader model to download it as a POJO:

1. Import the h2o module and spin up your H2O server:

    ```
    library(h2o)
    h2o.init()
    ```

2. Import the dataset by passing the location of the dataset in your system. Execute the following command:

    ```
    data_frame <- h2o.importFile("Dataset/iris.data")
    ```

3. Set the feature and label names by executing the following commands:

```
label <- "C5"
features <- setdiff(names(data), label)
```

4. Trigger AutoML by passing the training dataset, the feature columns, and the label columns as parameters. Also, set max_models to 10 and the seed value to 5:

```
aml <- h2o.automl(x = features, y = label, training_frame
= data_frame, max_models=10, seed = 5)
```

5. Once training is finished and you have the leaderboard, you can access the leader model using aml@leaderboard. We can also download the leader model as a POJO by executing the following command:

```
h2o.download_pojo(aml@leaderboard, path="~/Downloads/",
jar_name="AutoMLModel")
```

This will start downloading the AutoMLModel model POJO to your device at the specified path.

Now that we know how we can extract a POJO model in the R programming language, let's see how we can do this in H2O Flow.

Downloading H2O models as POJOs in H2O Flow

Downloading model POJOs in H2O Flow is very easy. H2O allows models to be downloaded as POJOs by simply clicking on a button. In *Chapter 2, Working with H2O Flow (H2O's Web UI)*, in the *Working with Model Training Functions in H2O Flow* section, you learned how to access a specific model's information.

For every model's information output in H2O Flow, in the **Actions** subsection, you have an interactive button titled **Download POJO**, as shown in the following screenshot:

Figure 10.1 – Gathering model information with the Download POJO button

You can simply click the **Download POJO** button to download the model as a POJO. You can download all the models that have been trained by H2O using this interactive button in H2O Flow.

Now that we have explored how we can download models as POJOs in Python, R, and H2O Flow, let's learn how to use this model POJO to make predictions.

Using a H2O model as a POJO

As mentioned in the previous section, a model POJO can be used on any platform that has a JVM installed. The only dependency is the h2o-genmodel.jar file, a JAR file that's needed to compile and run the model POJO to make predictions.

So, let's complete an experiment where we can use the model POJO along with the h2o-genmodel.jar file to understand how we can use model POJOs in any environment with JVM. We shall write a Java program that imports the h2o-genmodel.jar file and uses it to load the model POJO into the program. Once the model POJO has been loaded, we will use it to make predictions on the sample data.

So, let's start by creating a folder where we can keep the H2O POJO file needed for the experiment and then write some code that uses it. Follow these steps:

1. Open your terminal and create an empty folder by executing the following command:

    ```
    mkdir H2O_POJO
    cd H2O_POJO
    ```

2. Now, copy your model POJO file to the folder by executing the following command:

    ```
    mv {path_to_download_location}/{name_of_model_POJO} .
    ```

    Keep in mind that you may need to mention the name of the model you downloaded, as well as the path where you have downloaded your model POJO file.

3. Then, you need to download the h2o-genmodel.jar file. There are two ways you can do this:

    I. You can download the h2o-genmodel.jar file from your currently running local H2O server by running the following command:

        ```
        curl http://localhost:54321/3/h2o-genmodel.jar >
        h2o-genmodel.jar
        ```

    Keep in mind you will need an actively running H2O server present on localhost:54321. If your server is running on a different port, then edit the command with the appropriate port number.

    II. The h2o-genmodel.jar file is also available as a **Maven** dependency if you plan to use it in a Maven project. Apache Maven is a project management tool that does automated dependency management. Just add the following lines of code to your Maven pom.xml file inside its dependencies tag with, preferably, the latest version:

```
<dependency>
<dependency>
        <groupId>ai.h2o</groupId>
        <artifactId>h2o-genmodel</artifactId>
        <version>3.35.0.2</version>
</dependency>
```

The Maven repository for this can be found here: `https://mvnrepository.com/artifact/ai.h2o/h2o-genmodel`.

4. Now, let's create a sample Java program that uses the model POJO and the `h2o-genmodel.jar` file to make predictions on random data values. Create a Java program called `main.java` by executing the following command in your terminal:

    `vim main.java`

    This should open the `vim` editor for you to write your program in.

5. Let's start writing our Java program:

    I.   First, import the necessary dependencies, as follows:

    ```
    import hex.genmodel.easy.RowData;
    import hex.genmodel.easy.EasyPredictModelWrapper;
    import hex.genmodel.easy.prediction.*;
    ```

    II.   Then, create the `main` class, as follows:

    ```
    public class main { }
    ```

    III. Inside the `main` class, declare our model POJO's class name, as follows:

    ```
    private static final String modelPOJOClassName = "{name_
    of_model_POJO}";
    ```

    IV.   Then, create a `main` function inside the `main` class, as follows:

    ```
    public static void main(String[] args) throws Exception {
    }
    ```

V. Inside this `main` function, declare the `rawModel` variable as a `GenModel` object and initialize it by creating it as an instance of your model POJO by passing `modelPOJOClassName`, as follows:

```
hex.genmodel.GenModel rawModel;
rawModel = (hex.genmodel.GenModel) Class.
forName(modelPOJOClassName).getDeclaredConstructor().
newInstance();
```

VI. Now, let's wrap this `rawModel` object in an `EasyPredictModelWrapper` class. This class comes with easy-to-use functions that will make it easy for us to make predictions. Add the following code to your file:

```
EasyPredictModelWrapper model = new
EasyPredictModelWrapper(rawModel);
```

VII. Now that we have our `modelPOJO` object loaded and wrapped in `EasyPredictModelWrapper`, let's create some sample data for making predictions. Since we are using a model trained using the Iris dataset, let's create a `RowData` that contains C1, C2, C3, and C4 as features and some appropriate values. Add the following code to your file:

```
RowData row = new RowData();
row.put("C1", 5.1);
row.put("C2", 3.5);
row.put("C3", 1.4);
row.put("C4", 0.2);
```

VIII. Now, we need to create a prediction handler object that we can use to store the prediction results. Since the Iris dataset is for a multinomial classification problem, we will create an appropriate multinomial prediction handler object, as follows:

```
MultinomialModelPrediction predictionResultHandler =
model.predictMultinomial(row);
```

For different types of problems, you will need to use the appropriate types of prediction handler objects. You can find more information about this at https://docs.h2o.ai/h2o/latest-stable/h2o-genmodel/javadoc/index.html.

IX. Now, let's add some `print` statements so that we can get a clean and easy-to-understand output. Add the following `print` statements:

```
System.out.println("Predicted Class of Iris flower is: "
+ predictionResultHandler.label);
```

`predictionResultHandler.label` will contain the predicted label value.

X.   Let's also print out the different class probabilities so that we have an idea of what probability the label was predicted:

```
System.out.println("Class probabilities are: ");


for (int labelClassIndex = 0; labelClassIndex <
predictionResultHandler.classProbabilities.length;
labelClassIndex++) {


        System.out.println(predictionResultHandler.
classProbabilities[labelClassIndex]);


}
```

XI.   Finally, as the most important step, make sure all your braces are closed correctly and save the file.

6.   Once your file is ready, just compile the file by executing the following command:

```
javac -cp h2o-genmodel.jar -J-Xmx2g
-J-XX:MaxPermSize=128m DRF_1_AutoML_1_20220619_210236.
java main.java
```

7.   Once compilation is successful, execute the compiled file by running the following command in your Terminal:

```
java -cp .:h2o-genmodel.jar main
```

You should get the following output:

```
Predicted Class of Iris flower is: Iris-setosa
Class probabilities are:
0.9975669100660882
0.0
0.002433089933911768

Process finished with exit code 0
```

Figure 10.2 – Prediction results from the H2O model POJO implementation

As you can see, using the model POJO is very easy – you just need to create the POJO and use it in any regular Java program by implementing the `h2o-genmodel.jar` file.

> **Tip**
>
> If you plan on using model POJOs in production, then it is highly recommended that you understand the `h2o-genmodel.jar` library in detail. This library can provide you with lots of features and functionality that can make your deployment experience easy. You can find out more about this library here: `https://docs.h2o.ai/h2o/latest-stable/h2o-genmodel/javadoc/index.html`.

Congratulations! This chapter has helped you understand how to build, extract, and deploy model POJOs to make predictions on inbound data. You are now one step closer to using H2O in production.

Summary

In this chapter, we started by understanding what the usual problems are when working with an ML service in production. We understood how the portability of software, as well as ML models, plays an important role in seamless deployments. We also understood how Java's platform independence makes it good for deployments and how POJOs play a role in it.

Then, we explored what POJOs are and how they are independently functioning objects in the Java domain. We also learned that H2O has provisions to extract models trained by AutoML in the form of POJOs, which we can use as self-contained ML models capable of making predictions.

Building on top of this, we learned how to extract ML models in H2O as POJOs in Python, R, and H2O Flow. Once we understood how to download H2O ML models as POJOs, we learned how to use them to make predictions.

First, we understood that we need the `h2o-genmodel.jar` library and that it is responsible for interpreting the model POJO in Java. Then, we created an experiment where we downloaded the H2O model POJO and `h2o-genmodel.jar` and created a simple Java program that uses both of these files to make predictions on some sample data; this gave us some practical experience in working with model POJOs.

In the next chapter, we shall explore MOJOs, objects similar to POJOs but with some special benefits that can also be used in production.

11
Working with Model Object, Optimized (MOJO)

As we learned in *Chapter 10, Working with Plain Old Java Objects (POJOs)*, when working with production systems, we need portable software that we can easily deploy to our production servers. It is especially important in **Machine Learning (ML)** services that ML models be portable and self-sufficient. This helps engineers deploy new models regularly without worrying about breaking their systems in production because of any dependency issues.

H2O's model POJOs were a good solution to this problem. Model POJOs are H2O models that can be extracted in the form of Java POJOs that you can directly run using **Java Virtual Machine (JVM)** with the help of h2o-genmodel.jar.

However, model POJOs have certain drawbacks that prevent them from being the best solution to all these problems. When it comes to the portability of software packages, including POJOs, the smaller the object, the faster it is to deploy it. POJOs have an inherent limit on the size of the source files, which is up to 1 GB. Thus, models larger than 1 GB cannot be extracted as POJOs and at the same time, large models can be slow to deploy and perform.

That is why the team at H2O.ai created an alternative to POJOs called **Model Object, Optimized (MOJO)**. MOJOs are low-latency, self-sufficient, and standalone objects that can be easily deployed in production. They are smaller and faster counterparts to POJOs and are as easy to extract and use as POJOs.

In this chapter, we will cover the following topics:

- Understanding what a MOJO is
- Extracting H2O models as MOJOs
- Viewing model MOJOs
- Using H2O AutoML model MOJOs to make predictions

By the end of this chapter, you will be able to understand the difference between POJOs and MOJOs, extract trained models in the form of MOJOs using Python, R, or H2O Flow, and then use those MOJO models to load your ML program to make predictions.

Technical requirements

In this chapter, you will need the following:

- The latest version of your preferred web browser

- An **Integrated Development Environment** (**IDE**) of your choice

- (Optional) Jupyter Notebook by Project Jupyter (`https://jupyter.org/`)

  All the experiments conducted in this chapter are performed on a Terminal. You are free to follow along using the same setup or perform the same experiments using any IDE of your choice. All code examples for this chapter can be found on GitHub at `https://github.com/PacktPublishing/Practical-Automated-Machine-Learning-on-H2O/tree/main/Chapter%2011`.

Understanding what a MOJO is

MOJOs are counterparts to H2O model POJOs and technically work in the same way. H2O can build and extract models trained in the form of MOJOs, and you can use the extracted MOJOs to deploy and make predictions on inbound data.

So, what makes MOJOs different from POJOs?

POJOs have certain drawbacks that make them slightly less than ideal to use in a production environment, as follows:

- POJOs are not supported for source files larger than 1 GB, so any models with a size larger than 1 GB cannot be compiled to POJOs.

- POJOs do not support stacked ensemble models or Word2Vec models.

MOJOs, on the other hand, have the following additional benefits:

- MOJOs have no size restrictions

- MOJOs solve the large size issue by removing the ML tree and using a generic tree walking algorithm to navigate the model computationally

- MOJOs are smaller in size and faster than POJOs

- MOJOs support all types of models trained using H2O AutoML

As per H2O's in-house experiments and testing, as stated at `https://docs.h2o.ai/h2o/` `latest-stable/h2o-docs/productionizing.html#benefits-of-mojos-over-` `pojos`, it was noticed that MOJO models are roughly 20-25 times smaller in disk space than the respective POJO models. MOJOs were also twice as fast as POJOs when **hot scoring**, which is when scoring is done after the JVM has been able to optimize the execution paths. During **cold scoring**, which is when scoring is done before the JVM has optimized the execution path, MOJOs showed around 10-40 times faster execution compared to POJOs. MOJOs are more efficient compared to POJOs as the size of the model increases.

H2O's in-house testing also showed that when tested with 5,000 trees with a depth of 25, POJOs performed better when running binomial classification on very small trees of around 50 with 5 depths, but MOJOs performed better for multinomial classification.

Now that we know what MOJOs are, as well as their benefits, let's see how we can extract models trained using H2O's AutoML as MOJOs using simple examples.

Extracting H2O models as MOJOs

Just like POJOs, you can extract models trained using H2O's AutoML using any of the H2O-supported languages.

In the following sub-sections, we shall learn how to extract the model MOJOs using the Python and R programming languages, as well as see how we can extract model MOJOs using H2O Flow.

Extracting H2O models as MOJOs in Python

Let's see how we can extract models as MOJOs using Python. We shall use the same **Iris flower dataset** for running AutoML.

Follow these steps to train models using H2O AutoML. Then, we shall extract the leader model and download it as a MOJO:

1. Import the h2o module and spin up your H2O server:

    ```
    import h2o
    h2o.init()
    ```

2. Import the Iris dataset by passing the appropriate location of the dataset in your system. Execute the following command:

    ```
    data_frame = h2o.import_file("Dataset/iris.data")
    ```

3. Set the feature and label names by executing the following command:

```
features = data_frame.columns
label = "C5"
features.remove(label)
```

4. Initialize the H2O AutoML object by setting the value of the max_model parameter to 10 and the seed value to 5 by executing the following commands:

```
aml=h2o.automl.H2OAutoML(max_models=10, seed = 5)
```

5. Start the AutoML process by passing the training dataset, the feature columns, and the label column as parameters, as follows:

```
aml.train(x = features, y = label, training_frame = data_
frame)
```

6. Once training has finished, you can view the AutoML leaderboard by executing the following command:

```
print(aml.leaderboard)
```

You should get the following leaderboard:

model_id	mean_per_class_error	logloss	rmse	mse
GLM_1_AutoML_4_20220801_225630	0.0266667	0.067412	0.148337	0.0220039
GBM_lr_annealing_selection_AutoML_4_20220801_225630_select_model	0.0333333	0.140917	0.183485	0.0336669
GBM_grid_1_AutoML_4_20220801_225630_model_3	0.0466667	0.164239	0.206905	0.0428097
DRF_1_AutoML_4_20220801_225630	0.0533333	0.121106	0.193738	0.0375344
GBM_2_AutoML_4_20220801_225630	0.0533333	0.162145	0.210702	0.0443953
XRT_1_AutoML_4_20220801_225630	0.0533333	0.123873	0.197378	0.0389581
GBM_5_AutoML_4_20220801_225630	0.06	0.203821	0.224769	0.0505213
GBM_3_AutoML_4_20220801_225630	0.06	0.165065	0.210117	0.0441491
GBM_4_AutoML_4_20220801_225630	0.0666667	0.177578	0.221646	0.0491269

Figure 11.1 – AutoML leaderboard for extracting MOJOs

7. You can use aml.leader to get the leader model of the AutoML training. All models have an inbuilt function, download_mojo(), that extracts and downloads the model MOJO file:

```
aml.leader.download_mojo()
```

This should download the model MOJO to your device. You can also download a specific model from the leaderboard using `model_id`. Let's download the DRF model, which is ranked fourth on the leaderboard. Execute the following command:

```
DRF_model = h2o.get_model(aml.leaderboard[3,0])
DRF_model.download_mojo()
```

You can also specify the path where you want the MOJO file to be downloaded by passing the `path` parameter, along with the location, to the `download_mojo()` function. You can also download `h2o-genmodel.jar`, along with the MOJO file, by passing `get_genmodel_jar` as `True` in the `download_mojo()` function.

Let's see how we can do the same in the R programming language.

Extracting H2O models as MOJOs in R

Similar to how we can extract a model from the AutoML leaderboard in Python, we can do the same in the R programming language. We shall use the same Iris flower dataset again to train models using H2O AutoML and then extract the leader model to download it as a POJO. Follow these steps:

1. Import the h2o module and spin up your H2O server:

    ```
    library(h2o)
    h2o.init()
    ```

2. Import the dataset by passing the location of the dataset in your system. Execute the following command:

    ```
    data_frame <- h2o.importFile("Dataset/iris.data")
    ```

3. Set the feature and label names by executing the following commands:

    ```
    label <- "C5"
    features <- setdiff(names(data), label)
    ```

4. Trigger AutoML by passing the training dataset, the feature columns, and the label columns as parameters. Also, set `max_models` to `10` and the `seed` value to `5`:

    ```
    aml <- h2o.automl(x = features, y = label, training_frame
    = data_frame, max_models=10, seed = 5)
    ```

5. Once training has finished and you have the leaderboard, you can access the leader model using `aml@leaderboard`. Using the same command, we can download the leader model as a MOJO, like so:

    ```
    h2o.download_pojo(aml@leaderboard)
    ```

This will start downloading the model MOJO ZIP file to your device. Similar to Python, in R, you can specify the download path as well as set the `get_genmodel_jar` parameter to *True* to download the `h2o-genmodel.jar` file, along with the MOJO ZIP file.

Now that we know how to extract a model MOJO in the R programming language, let's learn how to do the same in H2O Flow.

Extracting H2O models as MOJOs in H2O Flow

Downloading model MOJOs in H2O Flow is just as easy as it was with POJOs. Right beside the **Download POJO** button, you have another button to download MOJO models.

As you learned in *Chapter 2, Working with H2O Flow (H2O's Web UI)*, in the *Working with Model Training Functions in H2O Flow* section, you can access specific model information.

In the **Actions** subsection, you have an interactive button titled **Model Deployment Package (MOJO)**, as shown in the following screenshot:

Figure 11.2 – The Download Model Deployment Package (MOJO) button

Simply clicking on this button will download the model as a MOJO. All models can be downloaded this way by using this interactive button in H2O Flow.

Unlike POJOs, where you have a single Java file, MOJOs can be downloaded as **ZIP files** that contain a collection of certain **configuration settings**, among other files. You can extract and explore these files if you wish, but from an implementation perspective, we will be using the whole ZIP file and using it in our services.

But regardless of the difference between the type of file, whether it is a Java file or a ZIP file, `h2o-genmodel.jar` has interpreters and readers for both types of files that you can use to read the models and make predictions.

Now that we have extracted the model MOJO, let's explore a special feature in MOJOs where we can graphically view the contents of a trained model.

Viewing model MOJOs

You can view MOJO models as simple human-readable graphs by using a Java tool called **Graphviz**. Graphviz is a piece of visualization software that is used for graphically visualizing structural information in the form of diagrams or graphs. It is a handy tool that is often used to show technical details in networks, web designs, and ML as simple images.

You can install the Graphviz library in different operating systems, as follows:

- **Linux**: You can download the library by just running the following command in your Terminal:

  ```
  sudo apt install graphviz
  ```

- **Mac**: You can use `brew` to install this library in your Mac system. Execute the following command in your Mac Terminal:

  ```
  brew install graphviz
  ```

- **Windows**: Graphviz has a Windows installer that you can download from `http://www.graphviz.org/download/`.

Once you have installed Graphviz, you can view the model graphically by using the `PrintMojo` function from the Terminal to make a **PNG file**.

Let's try it out. Execute the following steps:

1. Once you have downloaded your model MOJO file and installed Graphviz, you will need the `h2o.jar` file to be in the same path to access the `printMojo()` function in the `hex` class. You can download the `h2o.jar` file from `http://h2o-release.s3.amazonaws.com/h2o/rel-zumbo/2/index.html`.

2. Once your files are ready, open your Terminal in the same directory and execute the following command:

   ```
   java -cp h2o.jar hex.genmodel.tools.PrintMojo --tree 0 -i
   "DRF_1_AutoML_4_20220801_225630.zip" -o model.gv -f 20 -d
   3
   ```

   We are using the DRF model we downloaded from the experiment we did in the *Extracting H2O model as MOJO in Python* section. This command generates a `model.gv` file that the Graphviz visualization tool can use to visualize the model.

3. Now, use the Graphviz tool to build a PNG file using the `model.gv` file. Execute the following code:

   ```
   dot -Tpng model.gv -o model.png
   ```

   This generates the `model.png` file.

4. Now, open the `model.png` file; you should see an image of the model. The model should look as follows:

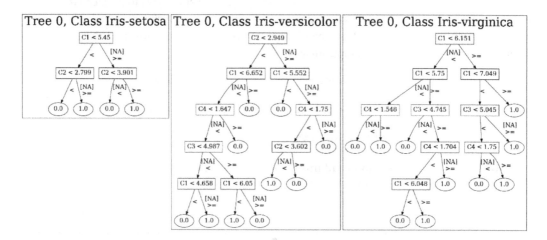

Figure 11.3 – Model image generated from MOJO using Graphviz

The preceding diagram is a nice graphical representation of how the decision tree of the **Distributed Random Forest (DRF)** model branches. You can also generate the model image directly with the `PrintMojo` function without needing the Graphviz library. However, this option is only available in Java 8 and higher versions.

5. Let's try using the `PrintMojo` function to generate the model image. Execute the steps in sequence to generate an image of the model without using Graphviz.

6. Similar to the previous experiment, where we printed the model MOJO using Graphviz, make sure that you have downloaded the model MOJO and copied it into a directory, along with your `h2o.jar` file. Now, open a Terminal in the same folder and execute the following command:

```
java -cp h2o.jar hex.genmodel.tools.PrintMojo --tree 0 -i
"DRF_1_AutoML_7_20220622_170835.zip" -o tree.png --format
png
```

The output of this command should generate a `tree.png` folder with images of the decision tree inside it. The diagram should look as follows:

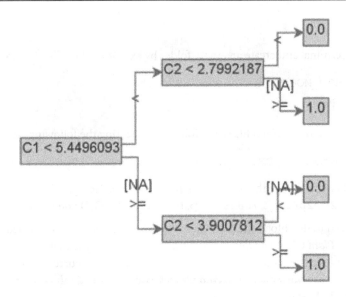

Figure 11.4 – Graphical images for class Iris-setosa using PrintMojo without Graphviz

Since we were using an ML model that has been trained on the Iris dataset, we have a multinomial classification model. Thus, inside the `tree.png` file, you will have individual images for every class – one for `Iris-setosa`, one for `Iris-virginica`, and one for `Iris-versicolor`.

Note that this feature is only available for tree-based algorithms such as DRF, GBM, and XGBoost. Linear models such as GLM and deep learning models are not supported for viewing.

Now that we know how to view the models from model MOJOs, let's learn how to use MOJOs to make predictions.

Using H2O AutoML model MOJOs to make predictions

Making predictions using MOJOs is the same as how we make predictions using model POJOS, albeit with some minor changes. Similar to POJOs, there is a dependency on the `h2o-genmodel.jar` file to compile and run the model MOJO to make predictions.

So, let's go ahead and quickly run an experiment where we can use the model MOJO with the `h2o-genmodel.jar` file to make predictions. We shall write a Java program that imports the `h2o-genmodel.jar` file and uses its classes to load and use our model MOJO to make predictions.

So, let's start by creating a folder where we can keep the H2O MOJO file needed for the experiment and then write some code that uses it.

Follow these steps:

1. Open your Terminal and create an empty folder by executing the following command:

    ```
    mkdir H2O_MOJO
    cd H2O_MOJO
    ```

2. Now, copy your model MOJO file to the folder by executing the following command:

    ```
    mv ~/Downloads/DRF_1_AutoML_7_20220622_170835.zip .
    ```

    Make sure that you change the name of the model MOJO, DRF_1_
    AutoML_7_20220622_170835.zip, to the model MOJO you are using.

3. Then, you need to download the h2o-genmodel.jar file. As you learned in *Chapter 10*,
    Working with Plain Old Java Objects (POJOs), there are two ways that you can do this. Either
    you can download the h2o-genmodel.jar file from your currently running local H2O
    server or, if you are working on a **Maven** project, you can just add the Maven dependency for
    h2o-genmodel, as follows:

    ```
    <dependency>

            <groupId>ai.h2o</groupId>

            <artifactId>h2o-genmodel</artifactId>

            <version>3.35.0.2</version>

    </dependency>
    ```

    The Maven repository for this can be found here: https://mvnrepository.com/
    artifact/ai.h2o/h2o-genmodel.

4. Now, let's create the Java program that will be making predictions using the model MOJO. Create
    a Java program called main.java by executing the following command in your Terminal:

    ```
    vim main.java
    ```

    This should open the vim editor, where you can write your code.

5. Let's start writing our Java program:

I. First, import the necessary dependencies, as follows:

```
import hex.genmodel.easy.RowData;

import hex.genmodel.easy.EasyPredictModelWrapper;

import hex.genmodel.easy.prediction.*;

import hex.genmodel.MojoModel;
```

II. Then, create the `main` class, as follows:

```
public class main { }
```

III. Then, inside the `main` class, create a `main` function, as follows:

```
public static void main(String[] args) throws Exception {
}
```

IV. Inside this `main` function, create the `EasyPredictModelWrapper` object by loading the MOJO model using the `MojoModel.load()` function and passing the location of your model MOJO. The code for this is as follows:

```
EasyPredictModelWrapper modelMOJO = new
EasyPredictModelWrapper(MojoModel.load("DRF_1_
AutoML_7_20220622_170835.zip"));
```

V. Now that we have our model MOJO loaded and wrapped in `EasyPredictModelWrapper`, let's create the sample data that we will use for making predictions. Add the following code to your file:

```
RowData row = new RowData();
row.put("C1", 5.1);
row.put("C2", 3.5);
row.put("C3", 1.4);
row.put("C4", 0.2);
```

VI.   Similar to how we did when making predictions using model POJOs, we need a prediction handler to store the prediction results from the model MOJOs. The prediction handler that's used for POJOs also works with MOJOs. So, let's create an appropriate multinomial prediction handler object, as follows:

```
MultinomialModelPrediction predictionResultHandler =
modelMOJO.predictMultinomial(row);
```

VII. Now, let's add the necessary `print` statements so that we have a clean and easy way to understand the output. Add the following `print` statements:

```
System.out.println("Predicted Class of Iris flower is: "
+ predictionResultHandler.label);
```

`predictionResultHandler.label` will contain the predicted label value.

VIII. Let's also print out the different class probabilities. Add the following code:

```
System.out.println("Class probabilities are: ");


for (int labelClassIndex = 0; labelClassIndex <
predictionResultHandler.classProbabilities.length;
labelClassIndex++) {


        System.out.println(predictionResultHandler.
classProbabilities[labelClassIndex]);


}
```

IX.   Make sure all your braces are closed correctly and save the file.

6.   Once your file is ready, just compile the file by executing the following command:

```
javac -cp h2o-genmodel.jar -J-Xmx2g
-J-XX:MaxPermSize=128m main.java
```

7.   Once the compilation is successful, execute the compiled file by running the following command in your Terminal:

```
java -cp .:h2o-genmodel.jar main
```

You should get the following output:

```
Predicted Class of Iris flower is: Iris-setosa
Class probabilities are:
0.9971308697337545
0.0
0.0028691302662455114

Process finished with exit code 0
```

Figure 11.5 – Prediction results from the H2O model MOJO implementation

As you can see, using the model MOJO is just as easy as using POJOs. Both are easy to extract and use in production. However, MOJOs benefit from being smaller and faster for large-sized models, which gives them a slight edge compared to POJOs.

Congratulations! You now know how to build, extract, and deploy model MOJOs to make predictions.

Summary

In this chapter, we started by understanding what the drawbacks of POJOs are. Then, we learned that H2O created a counterpart to POJOs called MOJOs, which do not have the same issues that POJOs have. Then, we learned what MOJOs are and the benefits of using them over POJOs. We learned that MOJOs are smaller and faster than POJOs. In H2O's internal experimentation, it was found that MOJOs performed better when working with large ML models.

After that, we learned how to practically extract ML models trained using AutoML as MOJOs. We understood how MOJOs can be downloaded in Python, R, and H2O Flow. Another benefit that we came across with MOJOs was that there is a special function called `PrintMojo` that can be used to create graphical pictures of ML models that can be read by humans. This also makes understanding the contents of the ML model easy.

Building on top of this knowledge, we implemented an experiment where we used the `h2o-genmodel.jar` file, along with the model MOJO, to make predictions on sample data, thus helping us get a better understanding of how we can use MOJOs in production.

In the next chapter, we shall explore the various design patterns that we can use to implement H2O AutoML. This will help us understand how to implement ideal ML solutions using H2O AutoML.

12
Working with H2O AutoML and Apache Spark

In *Chapter 10, Working with Plain Old Java Objects (POJOs)*, and *Chapter 11, Working with Model Object, Optimized (MOJO)*, we explored how to build and deploy our **Machine Learning** (ML) models as POJOs and MOJOs in production systems and use them to make predictions. In the majority of real-world problems, you will often need to deploy your entire ML pipeline in production so that you can deploy as well as train models on the fly. Your system will also be gathering and storing new data that you can later use to retrain your models. In such a scenario, you will eventually need to integrate your H2O server into your business product and coordinate the ML effort.

Apache Spark is one of the more commonly used technologies in the domain of ML. It is an analytics engine used for large-scale data processing using cluster computing. It is completely open source and widely supported by the Apache Software Foundation.

Considering the popularity of Spark in the field of data processing, H2O.ai developed an elegant software solution that combines the benefits of both Spark and AutoML into a single one-stop solution for ML pipelines. This software product is called H2O Sparkling Water.

In this chapter, we will learn more about H2O Sparkling Water. First, we will understand what Spark is and how it works and then move on to understanding how H2O Sparkling Water operates H2O AutoML in conjunction with Spark to solve fast data processing needs.

In this chapter, we will cover the following topics:

- Exploring Apache Spark
- Exploring H2O Sparkling Water

By the end of this chapter, you should have a general idea of how we can incorporate H2O AI along with Apache Spark using H2O Sparkling Water and how you can benefit from the best of both these worlds.

Technical requirements

For this chapter, you will require the following:

- The latest version of your preferred web browser.
- An **Integrated Development Environment (IDE)** of your choice or a Terminal.
- All experiments conducted in this chapter have been performed on a Terminal. You are free to follow along using the same setup or perform the same experiments using any IDE of your choice.

So, let's start by understanding what exactly Apache Spark is.

Exploring Apache Spark

Apache Spark started as a project in UC Berkeley AMPLab in 2009. It was then open sourced under a BSD license in 2010. Three years later, in 2013, it was donated to the Apache Software Foundation and became a top-level project. A year later, it was used by Databricks in a data sorting competition where it set a new world record. Ever since then, it has been picked up and used widely for in-memory distributed data analysis in the big data industry.

Let's see what the various components of Apache Spark are and their respective functionalities.

Understanding the components of Apache Spark

Apache Spark is an open source data processing engine. It is used to process data in real time, as well as in batches using cluster computing. All data processing tasks are performed in memory, making task executions very fast. Apache Spark's data processing capabilities coupled with H2O's AutoML functionality can make your ML system perform more efficiently and powerfully. But before we dive deep into H2O Sparkling Water, let's understand what Apache Spark is and what it consists of.

Let's start by understanding what the various components of the Spark ecosystem are:

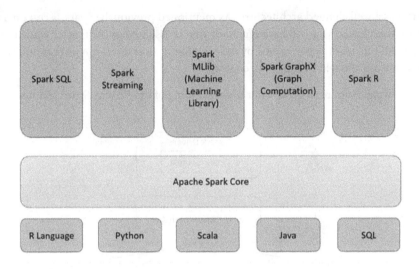

Figure 12.1 – Apache Spark components

The various components of the Spark ecosystem are as follows:

- **Spark Core**: The Spark Core component is the most vital component of the Spark ecosystem. It is responsible for basic functions such as input-output operations and scheduling and monitoring jobs. All the other components are built on top of this component. This component supports the Scala, Java, Python, and R programming languages using specific interfaces. The Spark Core component itself is written in the Scala programming language.

- **Spark SQL**: The Spark SQL component is used to leverage the power of SQL queries to run data queries on data stored in Spark nodes.

- **Spark Streaming**: The Spark Streaming component is used to batch as well as stream data in the same application.

- **Spark MLlib**: Spark MLlib is the ML library used by Spark to develop and deploy scalable ML pipelines. It is also used to perform ML analytics tasks such as feature extraction, feature engineering, dimensionality reduction, and so on.

- **GraphX**: The GraphX component is a library that is used to perform data analytics on graph-based data. It is used to perform graph data construction and traversals.

- **Spark R**: The Spark R component is an R package that provides a front-end shell for users to communicate with Spark via the R programming language. All data processing done by R is carried out on a single node. This makes R not ideal for processing large amounts of data. The Spark R component helps users perform these data operations on huge datasets in a distributed manner by using the underlying Spark cluster.

Understanding the Apache Spark architecture

Apache Spark has a well-defined architecture. As mentioned previously, Spark is run on a cluster system. Within this cluster, you will have one node that is assigned as the master while the others act as workers. All this work is performed by independent processes in the worker nodes and the combined effort is coordinated by the Spark context.

Refer to the following diagram to get a better understanding of the Apache Spark architecture:

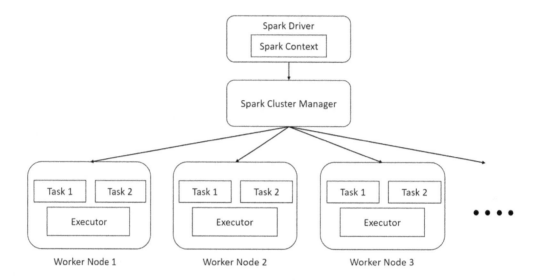

Figure 12.2 – Apache Spark architecture

The Spark architecture comprises the following components:

- **Spark Cluster Manager**: The Spark cluster manager is responsible for managing the allocation of resources to nodes and monitoring their health. It is responsible for maintaining the cluster of machines on which the Spark application is running. When you start a Spark application, the cluster manager will start up different nodes in the cluster, depending on the specified configuration, and restart any services that fail in the middle of execution.

  The Spark cluster manager comes in three types:

  - **Standalone**: This is a simple cluster manager that comes bundled with Spark and is very easy to set up and use.

- **Hadoop YARN: Yet Another Resource Negotiator (YARN)** is a resource manager that comes with the Hadoop ecosystem. Spark, being a data processing system, can integrate with many data storage systems. **Hadoop Distributed File System (HDFS)** is one of the most commonly used distributed filesystems in the big data industry and using Spark with HDFS has been a common setup in companies. Since YARN comes with the Hadoop ecosystem, you can use the same resource manager to manage your Spark resources.

- **Kubernetes**: Kubernetes is an open source container orchestration system for automating deployment operations, scaling services, and other forms of server management. Kubernetes is also capable of managing Spark cluster resources.

- **Spark Driver**: The Spark driver is the main program of the Spark application. It is responsible for controlling the execution of the application and keeps track of the different states of the nodes, as well as the tasks that have been assigned to each node. The program can be any script that you run or even the Spark interface.

- **Spark Executors**: The Spark executors are the actual processes that perform the computation task on the worker nodes. They are pretty simple processes whose aim is to take the assigned task, compute it, and then send back the results to the Spark Context.

- **SparkContext**: The Spark Context, as its name suggests, keeps track of the context of the execution. Any command that the Spark driver executes goes through this context. The Spark Context communicates with the Spark cluster manager to coordinate the execution activities with the correct executor.

The Spark driver program is the primary function that manages the parallel execution of operations on a cluster. The driver program does so using a data structure called a **Resilient Distributed Dataset (RDD)**.

Understanding what a Resilient Distributed Dataset is

Apache Spark is built on the foundation of the **RDD**. It is a fault-tolerant record of data that resides on multiple nodes and is immutable. Everything that you do in Spark is done using an RDD. Since it is immutable, any transformation that you do eventually creates a new RDD. RDDs are partitioned into logical sets that are then distributed among the Spark nodes for execution. Spark handles all this distribution internally.

Let's understand how Spark uses RDDs to perform data processing at scale. Refer to the following diagram:

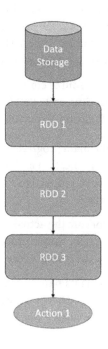

Figure 12.3 – Linear RDD transformations

So, RDDs are immutable, which means that once the dataset has been created, it cannot be modified. So, if you want to make changes in the dataset, then Spark will create a new RDD from the existing RDD and keeps track of the changes. Here, you have your initial data stored in **RDD 1**, so you must assume you need to drop a column and convert the type of another column from a string into a number. Spark will create **RDD 2**, which will contain these changes, as well as make note of the changes it has made. Eventually, as you further transform the data, Spark will contain many RDDs.

You may be wondering what happens if you need to perform many transformations on the data; will Spark create that many RRDs and eventually run out of memory? Remember, RDDs are resilient and immutable, so if you have created **RDD 3** from **RDD 2,** then you will only need to keep **RDD2** and the data transformation process from **RDD 2** to **RDD 3.** You will no longer need **RDD 1** so that can be removed to free up space. Spark does all the memory management for you. It will remove any RDDs that are not needed.

That was a very simplified explanation for a simple problem. What if you are creating multiple RDDs that contain different transformations from the same RDD? This can be seen in the following diagram:

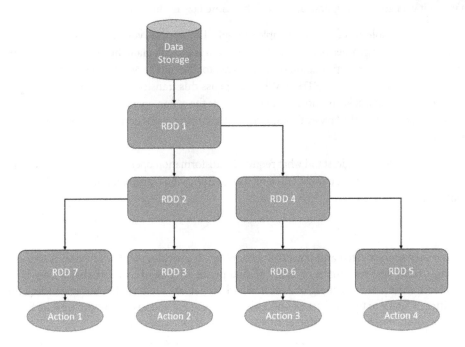

Figure 12.4 – Branched RDD transformations

In this case, you will need to keep all the RDDs. This is where Spark's **lazy evaluation** comes into play. Lazy evaluation is an evaluation technique where evaluation expressions are delayed until the resultant value is needed. Let's understand this better by looking into RDD operations. There are two types of operations:

- **Transformations**: Transformations are operations that produce a new RDD from an existing RDD that contains changes in the dataset. These operations mostly consist of converting raw datasets into a refined final dataset that can be used to extract evaluation metrics or other processes. This mostly involves data manipulation operations such as union operations or groupby operations.

- **Actions**: Actions are operations that take an RDD as input but don't generate a new RDD as output. The output value derived from the action operation is sent back to the driver. This mostly involves operations such as count, which returns the number of elements in the RDD, or aggregate, which performs aggregate operations on the contents of the RDD and sends the result back.

Transformation operations are lazy. When performing transformation operations on an RDD, Spark will keep a note of what needs to be done but won't do it immediately. It will only start the transformation process when it gets an action operation, hence the name lazy evaluation.

Let's understand the whole process with a simple example. Let's assume you have an RDD that contains a raw dataset of all the employees of a company and you want to calculate the average salary of all the senior ML engineers. Your transformation operations are to filter all the ML engineers into **RDD 2** and then further filter by seniority into **RDD 3** When you pass this transformation operation to Spark, it won't create **RDD 3** It will just keep a note of it. When it gets the action operation – that is, to calculate the average salary – that is when the lazy evaluation kicks in and Spark starts performing the transformation and, eventually, the action.

Lazy evaluation helps Spark understand what required transformation operations are needed to perform the action operation and find the most efficient way of doing the transformation while keeping the space complexity in mind.

> Tip
> Spark is a very sophisticated and powerful technology. It provides plenty of flexibility and can be configured to best suit different kinds of data processing needs. In this chapter, we have just explored the tip of the iceberg of Apache Spark. If you are interested in understanding the capabilities of Spark to their fullest extent, I highly encourage you to explore the Apache Spark documentation, which can be found at `https://spark.apache.org/`.

Now that we have a basic idea of how Spark works, let's understand how H2O Sparkling Water combines both H2O and Spark.

Exploring H2O Sparkling Water

Sparkling Water is an H2O product that combines the fast and scalable ML of H2O with the analytics capabilities of Apache Spark. The combination of both these technologies allows users to make SQL queries for data munging, feed the results to H2O for model training, build and deploy models to production, and then use them for predictions.

H2O Sparkling Water is designed in a way that you can run H2O in regular Spark applications. It has provisions to run the H2O server inside of Spark executors so that the H2O server has access to all the data stored in executors for performing any ML-based computations.

The transparent integration between H2O and Spark provides the following benefits:

- H2O algorithms, including AutoML, can be used in Spark workflows
- Application-specific data structures can be transformed and supported between H2O and Spark
- You can use Spark RDDs as datasets in H2O ML algorithms

Sparkling Water supports two types of backends:

- **Internal Backend**: In this type of setup, the H2O application is launched inside the Spark executor once the H2O context is initialized. H2O then starts its service by initializing its key-value store and memory manager inside each of the executors. It is easy to deploy H2O Sparkling Water as an internal backend, but if Spark's cluster manager decides to shut down any of the executors, then the H2O server running in the executor is also shut down. The internal backend is a default setup used by H2O Sparkling Water. The architecture of the internally running H2O Sparkling Water looks as follows:

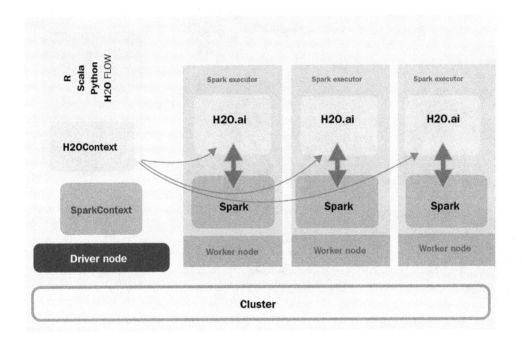

Figure 12.5 – Sparkling Water internal backend architecture

As you can see, the H2O service resides inside each of the Spark executors.

1. **External Backend**: In this type of setup, the H2O service is deployed separately from the Spark executors and the communication between the H2O servers and the Spark executors is handled by the Spark driver. The architecture of H2O Sparkling Water as an external backend works as follows:

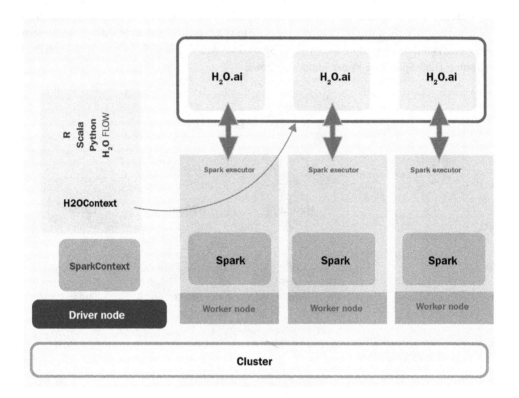

Figure 12.6 – Sparkling Water external backend architecture

As you can see, the H2O cluster is run separately from the Spark executor. The separation has benefits since the H2O clusters are no longer affected by the shutting down of Spark Executors. However, this adds the overhead of the H2O driver needing to coordinate the communication between the H2O cluster and the Spark Executors.

Sparkling Water, despite being built on top of Spark, uses an H2OFrame when performing computations using the H2O server in the Sparkling Water cluster. Thus, there is a lot of data exchange and interchange between the Spark RDD and the H2OFrame.

DataFrames are converted between different types as follows:

- **H2OFrame into RDD**: When converting an H2OFrame into an RDD, instead of recreating the data into a different type, Sparkling Water creates a wrapper around the H2OFrame that acts like an RDD API. This wrapper interprets all RDD-based operations into identical H2OFrame operations.

- **RDD into H2OFrame**: Converting an RDD into an H2OFrame involves evaluating the data in the RDD and then converting it into an H2OFrame. The data in the H2OFrame, however, is heavily compressed. Data being shared between H2O and Spark depends on the type of backend used for deployment.

- **Data Sharing in Internal Sparkling Water Backend**: In the internal Sparkling Water backend, since the H2O service is launched inside the Spark Executor, both the Spark service and the H2O service inside the executor use the same **Java Virtual Machine (JVM)** and as such, the data is accessible to both the services. The following diagram shows the process of data sharing in the internal Sparkling Water backend:

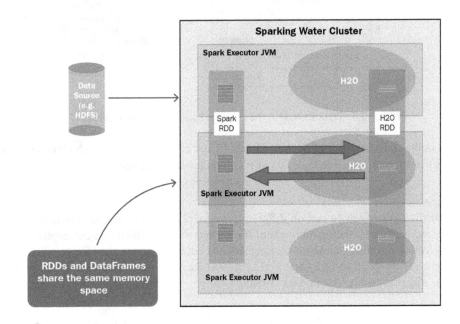

Figure 12.7 – Data sharing in the internal Sparkling Water backend

Since both services are on the same executor, you need to consider memory when converting DataFrames between the two types. You will need to allocate enough memory for both Spark and H2O to perform their respective operations. Spark will need the minimum memory of your dataset, plus additional memory for any transformations that you wish to perform. Also, converting RDDs into H2OFrames will lead to duplication of data, so it's recommended that a 4x bigger dataset should be used for H2O.

- **Data Sharing in External Sparkling Water Backend**: In the external Sparkling Water backend, the H2O service is launched in a cluster that is separate from the Spark Executor. So, there is an added overhead of transferring the data from one cluster to another over the network. The following diagram should help you understand this:

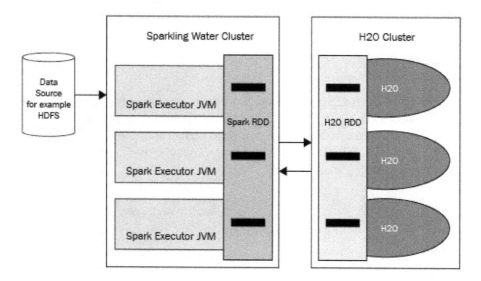

Figure 12.8 – Data sharing in the external Sparkling Water backend

Since both services reside in their own cluster (if you have allocated enough memory to the respective clusters), you don't need to worry about memory constraints.

> **Tip**
> H2O Sparkling Water can be run on different types of platforms in various kinds of ways. If you are interested in learning more about the various ways in which you can deploy H2O Sparkling Water, as well as getting more information about its backends, then feel free to check out https://docs.h2o.ai/sparkling-water/3.2/latest-stable/doc/design/supported_platforms.html.

Now that we know how H2O Sparkling Water works, let's see how we can download and install it.

Downloading and installing H2O Sparkling Water

H2O Sparkling Water has some specific requirements that need to be satisfied before you can install it on your system. The requirements for installing H2O Sparkling Water version 3.36 are as follows:

- **Operating System**: H2O Sparkling Water is only supported for Linux, macOS, and Windows.

- **Java Version**: H2O Sparkling Water supports all Java versions above Java 1.8.

- **Python Version**: If you plan to use the Python version of Sparkling Water, known as PySparkling, then you will need a Python version above 3.6 installed on your system.

- **H2O Version**: H2O Sparkling Water version 3.36.1 requires the same version of H2O installed on your system. However, H2O Sparkling Water comes prepackaged with a compatible H2O version, so you don't need to separately install H2O to use H2O Sparkling Water.

- **Spark Version**: H2O Sparkling Water version 3.36.1 strictly supports Spark 3.2. Any Spark version above or below version 3.2 may cause issues with installation or how H2O Sparkling Water works. Spark 3.2 has its own dependencies, which are as follows:

  - **Java Version**: Spark 3.2 strictly supports Java 8 and Java 11

  - **Scala Version**: Spark 3.2 strictly runs on Scala 2.12/2.13

  - **R Version**: Spark 3.2 supports any R version above 3.5

  - **Python Version**: Spark 3.2 supports any Python version above 3.6

- **Environment Variables**: You will need to set the SPARK_HOME environment variable to point to your local Spark 3.2 installation.

Now, let's set up our system so that we can download and install H2O Sparkling Water. Follow these steps to set up H2O Sparkling Water:

1. We will start by installing Java 11, which is needed for both Spark and H2O Sparkling Water. Even though Spark supports Java 8 as well, it is preferable to use Java 11 since it is the newer version with improvements and security patches. You can download and install Java 11 by executing the following command:

   ```
   sudo apt-get install openjdk-11-jdk
   ```

2. Optionally, if you wish to use the PySparkling Python interpreter, then install Python version 3.10. You can do so by executing the following command:

   ```
   sudo apt install python3
   ```

3. Now that we have the basic languages installed, let's go ahead and download and install Spark version 3.2. You can download the specific version for Spark from the Apache Software Foundation official download page (`https://www.apache.org/dyn/closer.lua/spark/spark-3.2.1/spark-3.2.1-bin-hadoop3.2.tgz`) or by directly running the following command in your Terminal:

    ```
    wget https://downloads.apache.org/spark/spark-3.1.2/
    spark-3.1.2-bin-hadoop3.2.tgz
    ```

    If you are using the **Maven project**, then you can directly specify the Spark core Maven dependency, as follows:

    ```
    <dependency>
        <groupId>org.apache.spark</groupId>
        <artifactId>spark-core_2.13</artifactId>
        <version>3.1.2</version>
    </dependency>
    ```

    You can find the Maven repository for Spark at `https://mvnrepository.com/artifact/org.apache.spark/spark-core`.

4. Then, you can extract the `.tar` file by executing the following command in your Terminal:

    ```
    sudo tar xzvf spark-*
    ```

5. Now that we have extracted the Spark binaries, let's set our environment variables, as follows:

    ```
    export SPARK_HOME="/path/to/spark/installation"
    ```

6. We must also set the `MASTER` environment variable to `local[*]` to launch a local Spark cluster:

    ```
    export MASTER="local[*]"
    ```

7. Now that we have all the dependencies of H2O Sparkling Water installed and ready, let's go ahead and download H2O Sparkling Water. You can download the latest version from `https://h2o.ai/products/h2o-sparkling-water/`. Upon clicking the **Download Latest** button, you should be redirected to the H2O Sparkling Water repository website, where you can download the H2O Sparkling Water version *3.36* ZIP file.

8. Once the download has finished, you can unzip the ZIP file by executing the following command in your Terminal:

    ```
    unzip sparkling-water-*
    ```

9. You can see if everything is working fine by starting the H2O Sparkling Water shell by executing the following command inside your Sparkling Water installation folder:

```
bin/sparkling-shell
```

10. By doing this, you can see if Sparkling Water has integrated with Spark by starting an H2O cloud inside the Spark cluster. You can do so by executing the following commands inside `sparkling-shell`:

```
import ai.h2o.sparkling._
val h2oContext = H2OContext.getOrCreate()
```

You should get an output similar to the following:

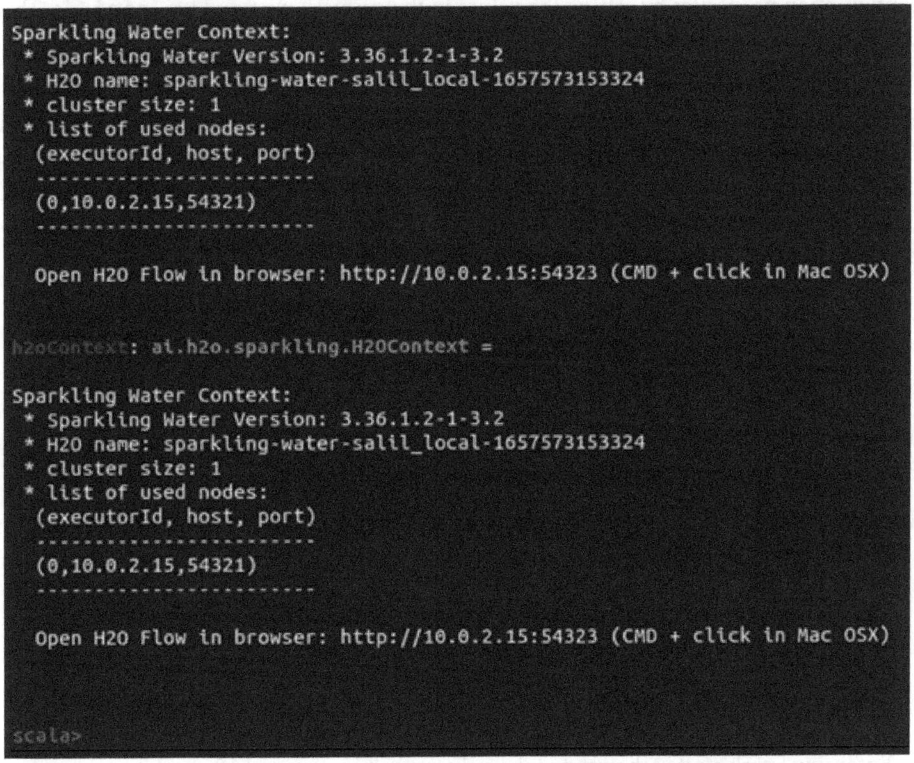

```
Sparkling Water Context:
 * Sparkling Water Version: 3.36.1.2-1-3.2
 * H2O name: sparkling-water-salil_local-1657573153324
 * cluster size: 1
 * list of used nodes:
   (executorId, host, port)
   ------------------------
   (0,10.0.2.15,54321)
   ------------------------

  Open H2O Flow in browser: http://10.0.2.15:54323 (CMD + click in Mac OSX)


h2oContext: ai.h2o.sparkling.H2OContext =

Sparkling Water Context:
 * Sparkling Water Version: 3.36.1.2-1-3.2
 * H2O name: sparkling-water-salil_local-1657573153324
 * cluster size: 1
 * list of used nodes:
   (executorId, host, port)
   ------------------------
   (0,10.0.2.15,54321)
   ------------------------

  Open H2O Flow in browser: http://10.0.2.15:54323 (CMD + click in Mac OSX)


scala>
```

Figure 12.9 – Successfully starting up H2O Sparkling Water

Now that we have successfully downloaded and installed both Spark and H2O Sparkling Water and ensured that both are working correctly, there is some general recommended tuning that you must do, as per H2O.ai's documentation. Let's take a look:

- Increase the available memory for the Spark driver as well as the Spark Executors from the default value of **1G** to **4G**. You can do so by passing the following `config` parameter when starting the Sparkling shell:

```
bin/sparkling-shell --conf spark.executor.memory=4g
spark.driver.memory=4g
```

    If you are using YARN or your cluster manager, then use `config spark.yarn.am.memory` instead of `spark.driver.memory`. You can also set these values as default configuration properties by setting the values in the `spark-defaults.conf` file. This can be found among your Spark installation files.

- Along with cluster memory, it is also recommended to increase the PermGen size of your Spark nodes. The default PermGen size often proves to be very small and can lead to `OutOfMemoryError`. **PermGen** is a special heap space that is separate from the main memory heap. It is used by the JVM to keep track of loaded class metadata. You can set this value using the `spark.driver.extraJavaOptions` and `spark.executor.extraJavaOptions` configuration options, as follows:

```
bin/sparkling-shell --conf spark.driver.extraJavaOptions
-XX:MaxPermSize=384 -XX:PermSize=384m spark.executor.
extraJavaOptions -XX:MaxPermSize=384 -XX:PermSize=384m
```

- It is also recommended to keep your cluster homogeneous – that is, both the Spark driver and Executor have the same amount of resources allocated to them.

- The following configurations are also recommended to speed up and stabilize the creation of H2O services on top of the Spark cluster:

    - Increase the number of seconds to wait for a task launched in data-local mode so that H2O tasks are processed locally with data. You can set this as follows:

```
bin/sparkling-shell --conf spark.locality.wait=3000
```

    - Enforcing Spark starts scheduling jobs only when it is allocated 100% of its resources:

```
bin/sparkling-shell --conf spark.scheduler.
minRegisteredResourcesRatio=1
```

    - Don't retry failed tasks:

```
bin/sparkling-shell --conf spark.task.maxFailures=1
```

- Set the interval between each executor's heartbeats to the driver to less than Spark's network timeout – that is `spark.network.timeout` – whose default value is *120 seconds*. So, set the heartbeat value to around *10 seconds*:

```
bin/sparkling-shell --conf spark.executor.
heartbeatInterval=10s
```

Now that we have appropriately configured Spark and H2O Sparkling Water, let's see how we can use these technologies to solve an ML problem using both Spark and H2O AutoML.

Implementing Spark and H2O AutoML using H2O Sparkling Water

For this experiment, we will be using the Concrete Compressive Strength dataset. You can find this dataset at `https://archive.ics.uci.edu/ml/datasets/Concrete+Compressive+Strength`.

Here are more details on the dataset: I-Cheng Yeh, *Modeling of strength of high performance concrete using artificial neural networks*, Cement and Concrete Research, Vol. 28, No. 12, pp. 1797-1808 (1998).

Let's start by understanding the problem statement we will be working with.

Understanding the problem statement

The Concrete Compressive Strength dataset is a dataset that consists of *1,030* data points consisting of the following features:

- **Cement**: This feature denotes the amount of cement added in kg in m3 of the mixture
- **Blast Furnace Slag**: This feature denotes the amount of slag added in kgs in m3 of the mixture
- **Fly Ash**: This feature denotes the amount of fly ash added in kgs in m3 of the mixture
- **Water**: This feature denotes the amount of water added in kgs in m3 of the mixture
- **Superplasticizer**: This feature denotes the amount of superplasticizer added in kgs in m3 of the mixture
- **Coarse Aggregate**: This feature denotes the amount of coarse aggregate – in other words, stone – added in kgs in m3 of the mixture
- **Fine Aggregate**: This feature denotes the amount of fine aggregate – in other words, sand – added in kgs in m3 of the mixture
- **Age**: This feature denotes the age of the cement
- **Concrete compressive strength**: This feature denotes the compressive strength of the concrete in **Megapascals (MPa)**

The ML problem is to use all the features to predict the compressive strength of the concrete.

The content of the dataset is as follows:

Cement	Blast Furnace Slag	Fly Ash	Water	Superplasticizer	Coarse Aggregate	Fine Aggregate	Age	Concrete Compressive Strength
540.0	0.0	0.0	162.0	2.5	1040.0	676.0	28	79.99
540.0	0.0	0.0	162.0	2.5	1055.0	676.0	28	61.89
332.5	142.5	0.0	228.0	0.0	932.0	594.0	270	40.27
332.5	142.5	0.0	228.0	0.0	932.0	594.0	365	41.05
540.0	0.0	0.0	162.0	2.5	1040.0	676.0	28	79.99
540.0	0.0	0.0	162.0	2.5	1055.0	676.0	28	61.89
...	...	...	...	...	...	...	...	...

Figure 12.10 – Concrete Compressive Strength dataset sample

So, let's see how we can solve this problem using H2O Sparkling Water. First, we shall learn how to train models using H2O AutoML and Spark.

Running AutoML training in Sparkling Water

Once you have successfully installed both Spark 3.2 and H2O Sparkling Water, as well as set the correct environment variables (SPARK_HOME and MASTER), you can start the model training process.

Follow these steps:

1. Start the Sparkling shell by executing the command inside the H2O Sparkling Water extracted folder:

    ```
    ./bin/sparkling-shell
    ```

This should start a Scala shell in your Terminal. The output should look as follows:

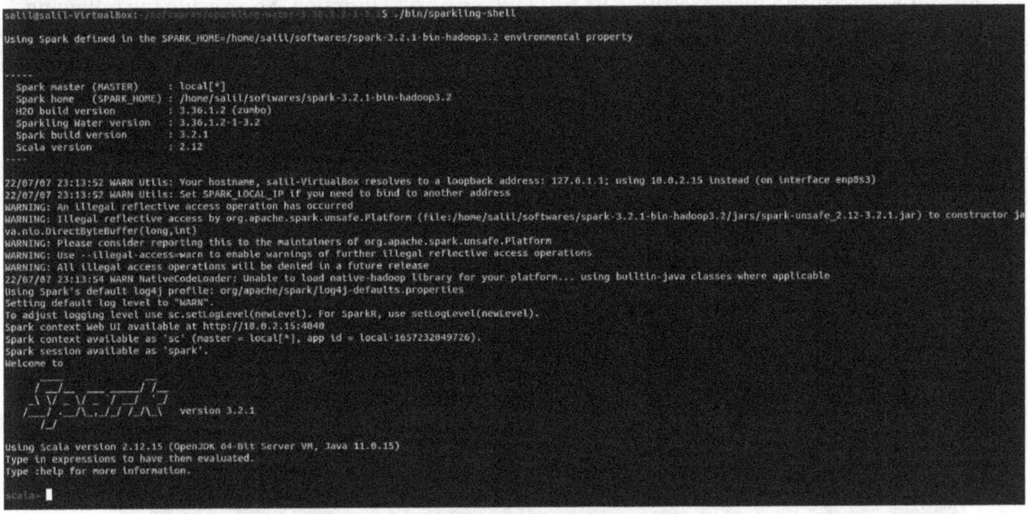

Figure 12.11 – Scala shell for H2O Sparkling Water

You can also perform the same experiment in Python using the `PySparkling` shell. You can start the `PySparkling` shell by executing the following command:

```
./bin/PySparkling
```

You should get an output similar to the following:

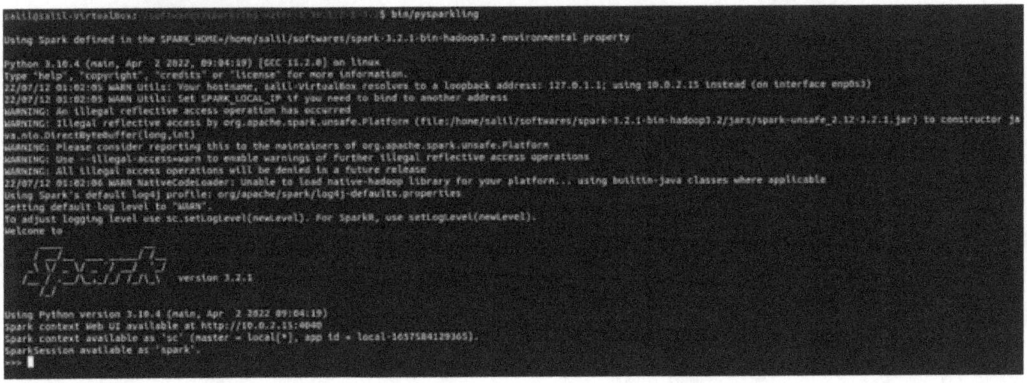

Figure 12.12 – Python shell for H2O Sparkling Water

2. Now, we need to start an H2O cluster inside the Spark environment. We can do this by creating an H2OContext and then executing its `getOrCreate()` function. So, execute the following code in your Sparkling shell to import the necessary dependencies and execute the H2O context code:

```
import ai.h2o.sparkling._
import java.net.URI
val h2oContext = H2OContext.getOrCreate()
```

In the PySparkling shell, the code will be as follows:

```
from PySparkling import *
h2oContext = H2OContext.getOrCreate()
```

You should get an output similar to the following that states that your H2O context has been created:

```
h2oContext: ai.h2o.sparkling.H2OContext =

Sparkling Water Context:
 * Sparkling Water Version: 3.36.1.2-1-3.2
 * H2O name: sparkling-water-salil_local-1657232049726
 * cluster size: 1
 * list of used nodes:
  (executorId, host, port)
  ------------------------
  (0,10.0.2.15,54321)
  ------------------------

  Open H2O Flow in browser: http://10.0.2.15:54323 (CMD + click in Mac OSX)
```

Figure 12.13 – H2O context created successfully

3. Now, we must ensure that our Concrete Compressive Strength dataset can be downloaded on every node using Spark's built-in file I/O system. So, execute the following commands to import your dataset:

```
import org.apache.spark.SparkFiles
spark.sparkContext.addFile("/home/salil/Downloads/
Concrete_Data.csv")
```

In the PySparkling shell, we must import the dataset using H2O's `import` function. The Python code will be as follows:

```
import h2o
h2oFrame = h2o.import_file("/home/salil/Downloads/
Concrete_Data.csv")
```

4. Once added, we must parse the dataset into a Spark Dataframe by executing the following commands in the Scala shell:

```
val sparkDataFrame = spark.read.option("header",
"true").option("inferSchema", "true").csv(SparkFiles.
get("Concrete_Data.csv"))
```

In the PySparkling shell, the equivalent code will be as follows:

```
sparkDataFrame = hc.asSparkFrame(h2oFrame)
```

5. Now, `sparkDataFrame` contains the dataset as a Spark DataFrame. So, let's perform a train-test split on it to split the DataFrame into testing and training DataFrames. You can do so by executing the following commands in the Sparkling shell:

```
val Array(trainingDataFrame, testingDataFrame) =
sparkDataFrame.randomSplit(Array(0.7, 0.3), seed=123)
```

In the PySparkling shell, execute the following command:

```
[trainingDataFrame, testingDataFrame] = sparkDataFrame.
randomSplit([0.7, 0.3], seed=123)
```

6. We now have `trainingDataFrame` and `testingDataFrame` ready for training and testing, respectively. Let's create an H2OAutoML instance to auto-train models on `trainingDataFrame`. Execute the following commands to instantiate an H2O AutoML object:

```
import ai.h2o.sparkling.ml.algos.H2OAutoML
val aml = new H2OAutoML()
```

In PySparkling, when initializing the H2O AutoML object, we also set the label column. The code for this is as follows:

```
from PySparkling.ml import H2OAutoML
aml = H2OAutoML(labelCol=" Concrete compressive strength
")
```

7. Let's see how we can set the label of the dataset so that the AutoML object is aware of which columns from the DataFrame are to be predicted in the Scala shell. Execute the following command:

```
aml.setLabelCol("Concrete compressive strength")
```

H2O will treat all the columns of the DataFrame as features unless explicitly specified. It will, however, ignore columns that are set as **labels**, **fold columns**, **weights**, or any other explicitly set ignored columns.

H2O AutoML distinguishes between regression and classification problems depending on the type of the response column. If the response column is a string, then H2O AutoML assumes it is a **classification problem**, whereas if the response column is numerical, then H2O AutoML assumes that it is a **regression problem**. You can override this default behavior by explicitly specifying this during instantiation by either instantiating the `ai.h2o.sparkling.ml.algos.classification.H2OAutoMLClassifier` object or the `ai.h2o.sparkling.ml.algos.regression.H2OAutoMLRegressor` object instead of `ai.h2o.sparkling.ml.algos.H2OautoML`, as we did in this example.

8. Now, let's limit AutoML model training to only 10 models. Execute the following command:

```
aml.setMaxModels(10)
```

The equivalent Python syntax for this code is the same, so execute this same command in your PySparkling shell.

9. Once we have our AutoML object all set up, the only thing remaining is to trigger the training. To do so, execute the following command:

```
val model = aml.fit(trainingDataFrame)
```

The equivalent code for Python is as follows:

```
model = aml.fit(trainingDataFrame)
```

Once training is finished, you should get an output similar to the following:

```
Returning leader model and printing info about it below.
Model Details
==============
H2OStackedEnsemble
Model Key: StackedEnsemble_AllModels_1_AutoML_2_20220709_155143_e0a6c5512727

Model summary

Training metrics
RMSLE: 0.06461275612484682
Nobs: 711.0
RMSE: 1.664316734598618
ResidualDeviance: 1969.4345872692202
NullDeviance: 201217.88337876744
MAE: 1.180679697214416
MeanResidualDeviance: 2.7699501930650072
ScoringTime: 1.657378357095E12
MSE: 2.7699501930650072
R2: 0.9902124276719378
NullDegreesOfFreedom: 710.0
AIC: 2760.118254384428
ResidualDegreesOfFreedom: 703.0

Cross validation metrics
RMSLE: 0.1460912336455651
Nobs: 711.0
RMSE: 4.324493096662628
ResidualDeviance: 13296.582026131815
NullDeviance: 201708.75554137037
MAE: 2.9346239510372185
MeanResidualDeviance: 18.701240543082722
ScoringTime: 1.657378356816E12
MSE: 18.701240543082722
R2: 0.9339194816938675
NullDegreesOfFreedom: 710.0
AIC: 4119.957985290266
ResidualDegreesOfFreedom: 702.0
```

Figure 12.14 – H2O AutoML result in H2O Sparkling Water

As you can see, we got a stacked ensemble model as the leader model with the model key below it. Below **Model Key** is **Model summary**, which contains the training and cross-validation metrics.

As we did in *Chapter 2, Working with H2O Flow (H2O's Web UI)*, we have not set the sort metric for the aml object, so by default, H2O AutoML will use the default metrics. This will be deviance since it is a **regression problem**. You can explicitly set the sort metric using automl.setSortMetric() and pass in the sort metric of your choice.

10. You can also get a detailed view of the model by using the `getModelDetails()` function. Execute the following command:

```
model.getModelDetails()
```

This command will work on both the PySparkling and Scala shells and will output very detailed JSON about the model's metadata.

11. You can also view the AutoML leaderboard by executing the following command:

```
val leaderboard = aml.getLeaderboard()
leaderboard.show(false)
```

The equivalent Python code for the PySparkling shell is as follows:

```
leaderboard = aml.getLeaderboard("ALL")
leaderboard.show(truncate = False)
```

You should get an output similar to the following:

	model_id	rmse	mse	mae	rmsle	mean_residual_deviance
0	StackedEnsemble_AllModels_1_AutoML_2_20220709_155143	4.324493096662628	18.701240543082722	2.9346239510372185	0.1460912336455651	18.701240543082722
1	StackedEnsemble_BestOfFamily_1_AutoML_2_20220709_155143	4.384260179975179	19.221737325715992	2.9686839361381865	0.1481222119829288	19.221737325715992
2	GBM_4_AutoML_2_20220709_155143	4.573584379622189	20.917674077524083	3.1189841752490692	0.1451041929152035	20.917674077524083
3	XGBoost_3_AutoML_2_20220709_155143	4.594722541302787	21.11147523155594	3.14733816048264	0.1695988536803904	21.11147523155594
4	GBM_3_AutoML_2_20220709_155143	4.602836781243628	21.186106434769197	3.216627946091281	0.1484644220905108	21.186106434769197
5	XGBoost_2_AutoML_2_20220709_155143	4.795345366837029	22.995337187245365	3.2627855644950383	0.163751109572822	22.995337187245365
6	GBM_2_AutoML_2_20220709_155143	4.805228403927745	23.09022001391398	3.357554479064718	0.1556970476685876	23.09022001391398
7	XGBoost_1_AutoML_2_20220709_155143	4.883657455766908	23.850110145267706	3.4043382679527485	0.1672261993978137	23.850110145267706
8	XRT_1_AutoML_2_20220709_155143	5.925199691510438	35.107991384275394	4.320396072254607	0.2074160378479092	35.107991384275394
9	DRF_1_AutoML_2_20220709_155143	5.946360562106173	35.35920393457163	4.299032302631552	0.2060104350171061	35.35920393457163
10	GBM_1_AutoML_2_20220709_155143	6.787876516135789	46.07526759830773	5.291743594342693	0.2203005198493777	46.07526759830773
11	GLM_1_AutoML_2_20220709_155143	10.520585472187229	110.68271867759698	8.317403167170115	0.3410797993409181	110.68271867759698

Figure 12.15 – H2O AutoML leaderboard in H2O Sparkling Water

This will display the leaderboard containing all the models that have been trained and ranked based on the sort metric.

12. Making predictions using H2O Sparkling Water is also very easy. The prediction functionality is wrapped behind a simple, easy-to-use wrapper function called `transform`. Execute the following code to make predictions on the testing DataFrame:

```
model.transform(testingDataFrame).show(false)
```

In the PySparkling shell, it is slightly different. Here, you must execute the following code:

```
model.transform(testingDataFrame).show(truncate = False)
```

You should get an output similar to the following:

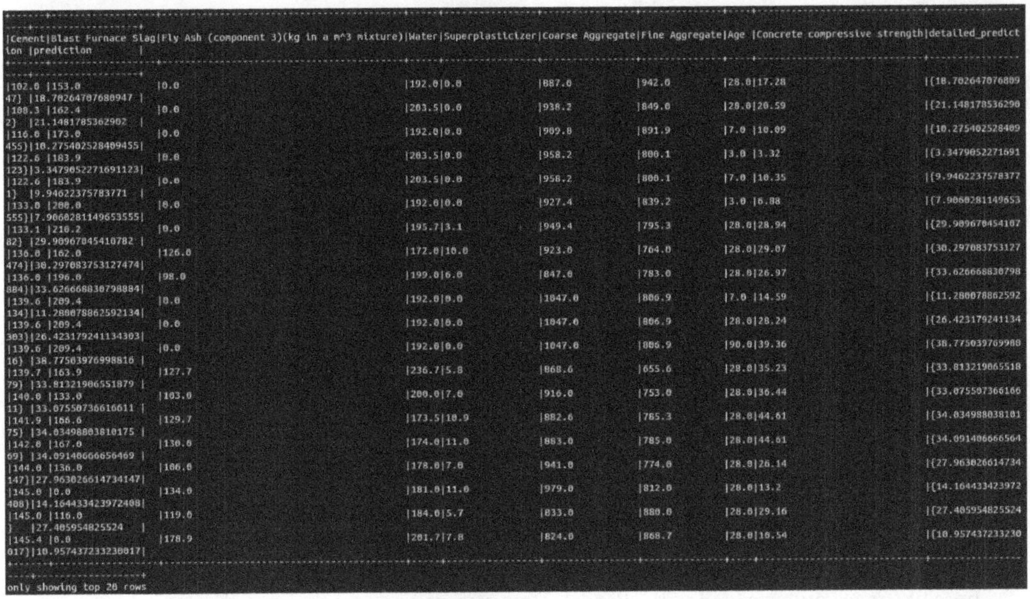

```
|Cement|Blast Furnace Slag|Fly Ash (component 3)(kg in a m^3 mixture)|Water|Superplasticizer|Coarse Aggregate|Fine Aggregate|Age |Concrete compressive strength|detailed_predict
ion |prediction          |
|102.0 |153.0             |0.0        |192.0|0.0        |887.0       |942.0        |28.0|17.28        |{18.702647076809
47} |18.70264707680947 |
|108.3 |162.4             |0.0        |203.5|0.0        |938.2       |849.0        |28.0|20.59        |{21.148178536290
2}  |21.1481785362902  |
|116.0 |173.0             |0.0        |192.0|0.0        |909.8       |891.9        |7.0 |10.09        |{10.275402528409
455}|10.275402528409455|
|122.6 |183.9             |0.0        |203.5|0.0        |958.2       |800.1        |3.0 |3.32         |{3.3479052271691
123}|3.3479052271691123|
|122.6 |183.9             |0.0        |203.5|0.0        |958.2       |800.1        |7.0 |10.35        |{9.9462237578377
1}  |9.94622375783771  |
|133.0 |200.0             |0.0        |192.0|0.0        |927.4       |839.2        |3.0 |6.88         |{7.9060281149653
555}|7.9060281149653555|
|133.1 |210.2             |0.0        |195.7|3.1        |949.4       |795.3        |28.0|28.94        |{29.909670454107
82} |29.90967045410782 |
|136.0 |162.0             |126.0      |172.0|10.0       |923.0       |764.0        |28.0|29.07        |{30.297083753127
474}|30.297083753127474|
|136.0 |196.0             |98.0       |199.0|6.0        |847.0       |783.0        |28.0|26.97        |{33.626668830798
884}|33.626668830798884|
|139.6 |209.4             |0.0        |192.0|0.0        |1047.0      |806.9        |7.0 |14.59        |{11.280078862592
134}|11.280078862592134|
|139.6 |209.4             |0.0        |192.0|0.0        |1047.0      |806.9        |28.0|28.24        |{26.423179241134
303}|26.423179241134303|
|139.6 |209.4             |0.0        |192.0|0.0        |1047.0      |806.9        |90.0|39.36        |{38.775039769988
16} |38.77503976998816 |
|139.7 |163.9             |127.7      |236.7|5.8        |868.6       |655.6        |28.0|35.23        |{33.813219065518
79} |33.81321906551879 |
|140.0 |133.0             |103.0      |200.0|7.0        |916.0       |753.0        |28.0|36.44        |{33.075507366166
11} |33.07550736616611 |
|141.9 |166.6             |129.7      |173.5|10.9       |882.6       |785.3        |28.0|44.61        |{34.034988038101
75} |34.03498803810175 |
|142.0 |167.0             |130.0      |174.0|11.0       |883.0       |785.0        |28.0|44.61        |{34.091406666564
69} |34.09140666656469 |
|144.0 |136.0             |106.0      |178.0|7.0        |941.0       |774.0        |28.0|26.14        |{27.963026614734
147}|27.963026614734147|
|145.0 |0.0               |134.0      |181.0|11.0       |979.0       |812.0        |28.0|13.2         |{14.164433423972
408}|14.164433423972408|
|145.0 |116.0             |119.0      |184.0|5.7        |833.0       |880.0        |28.0|29.16        |{27.405954825524
}   |27.405954825524   |
|145.4 |0.0               |178.9      |201.7|7.8        |824.0       |868.7        |28.0|10.54        |{10.957437233230
017}|10.957437233230017|
only showing top 20 rows
```

Figure 12.16 – Prediction results combined with the testing DataFrame

The output of the `transform` function shows the entire **testDataFrame** with two additional columns on the right-hand side called **detailed_prediction** and **prediction**.

13. Now, let's download this model as a MOJO so that we can use it for the next experiment, where we shall see how H2O Sparkling Water loads and uses MOJO models. Execute the following command:

```
model.write.save("model_dir")
```

This command is the same for both the Scala and Python shells and should download the model MOJO in your specified path. If you are using the Hadoop filesystem as your Spark data storage engine, then the command uses HDFS by default.

Now that we know how to import a dataset, train models, and make predictions using H2O Sparkling Water, let's take it one step further and see how we can reuse existing model binaries, also called MOJOs, by loading them into H2O Sparkling Water and making predictions on them.

Making predictions using model MOJOs in H2O Sparkling Water

When you train models using H2O Sparkling Water, the models that are generated are always of the MOJO type. H2O Sparkling Water can load model MOJOs generated by H2O-3 and is also backward compatible with the different versions of H2O-3. You do not need to create an H2O context to use model MOJOs for predictions, but you do need a scoring environment. Let's understand this by completing an experiment.

Follow these steps:

1. To make predictions using imported model MOJOs, you need a scoring environment. We can create a scoring environment in two ways; let's take a look:

    I. Use Sparkling Water prepared scripts, which set all the dependencies that are needed to load MOJOs and make predictions on the Spark classpath. Refer to the following commands:

    The following command is for a Scala shell:

    ```
    ./bin/spark-shell --jars jars/sparkling-water-assembly-
    scoring_2.12-3.36.1.3-1-3.2-all.jar
    ```

    The following command is for a Python shell:

    ```
    ./bin/pyspark --py-files py/h2o_PySparkling_scoring_3.2-
    3.36.1.3-1-3.2.zip
    ```

    II. Use Spark directly and set the dependencies manually.

2. Once we have our scoring environment set up, we can load the model MOJOs. Model MOJOs loaded into Sparkling Water are immutable. So, making any configuration changes is not possible once you have loaded the model. However, you can set the configurations before you load the model. You can do so by using the `H2OMOJOSettings()` function. Refer to the following example:

    ```
    import ai.h2o.sparkling.ml.models._
    val modelConfigurationSettings =
    H2OMOJOSettings(convertInvalidNumbersToNa = true,
    convertUnknownCategoricalLevelsToNa = true)
    ```

    For PySparkling, refer to the following code:

    ```
    from PySparkling.ml import *
    val modelConfigurationSettings =
    H2OMOJOSettings(convertInvalidNumbersToNa = true,
    convertUnknownCategoricalLevelsToNa = true)
    ```

3. Once you have set the configuration settings, you can load the model MOJO using the `createFromMojo()` function from the `H2OMOJOModel` library. So, execute the following code to load the model MOJO that you created in the *Running AutoML training in Sparkling Water* section and pass the configuration settings:

    ```
    val loadedModel = H2OMOJOModel.createFromMojo("model_dir/
    model_mojo", modelConfigurationSettings)
    ```

The Python equivalent is as follows:

```
loadedModel = H2OMOJOModel.createFromMojo("model_dir/
model_mojo", modelConfigurationSettings)
```

If you specify the model MOJO path as a relative path and if HDFS is enabled, Sparkling Water will check the HDFS home directory; otherwise, it will search for it from the current directory. You can also pass an absolute path to your model MOJO file.

You can also manually specify where you want to load your model MOJO. For the HDFS filesystem, you can use the following command:

```
loadedModel = H2OMOJOModel.createFromMojo("hdfs:///user/
salil/ model_mojo")
```

For a local filesystem, you can use the following command:

```
loadedModel = H2OMOJOModel.createFromMojo("file:///Users/
salil/some_ model_mojo")
```

4. Once successfully loaded, you can simply use the model to make predictions, as we did in the *Running AutoML training in Sparkling Water* section. So, execute the following command to make predictions using your recently loaded model MOJO:

```
val predictionResults = loadedModel.
transform(testingDataframe)
```

5. The prediction results are stored as another Spark DataFrame. So, to view the prediction values, we can just display the prediction results by executing the following command:

```
predictionResults.show()
```

You should get an output similar to the following:

```
|Cement|Blast Furnace Slag|Fly Ash|Water|Superplasticizer|Coarse Aggregate|Fine Aggregate| Age|Concrete compressive strength| detailed_prediction|          prediction|
| 102.0|             153.0|    0.0|192.0|             0.0|           887.0|         942.0|28.0|                        17.28| [19.20426780893517]| 19.20426780893517|
| 108.3|             162.4|    0.0|203.5|             0.0|           938.2|         849.0|28.0|                        28.59| [21.78240293412489]| 21.78240293412489|
| 116.0|             173.0|    0.0|192.0|             0.0|           909.8|         891.9| 7.0|                        10.09| [9.418119576417453]| 9.418119576417453|
| 122.0|             183.9|    0.0|203.5|             0.0|           958.2|         800.1| 3.0|                         3.32| [4.788849427699202]| 4.788849427699202|
| 122.6|             183.9|    0.0|203.5|             0.0|           958.2|         800.1| 7.0|                        10.35|[10.959896897256051]|10.959896897256051|
| 133.0|             200.0|    0.0|192.0|             0.0|           927.4|         839.2| 3.0|                         0.88| [8.052752331426289]| 8.052752331426289|
| 133.1|             210.2|    0.0|195.7|             3.1|           949.4|         795.3|28.0|                        28.94|[30.209503295883003]|30.209503295883003|
| 136.0|             162.0|  126.0|172.0|            10.0|           923.0|         764.0|28.0|                        29.07| [31.24308884928773]| 31.24308884928773|
| 136.0|             196.0|   98.0|199.0|             6.0|           847.0|         783.0|28.0|                        26.97|[34.16633337130581]| 34.16633337130581|
| 139.6|             209.4|    0.0|192.0|             0.0|          1047.0|         806.9| 7.0|                        14.59|[12.15639740938616]| 12.15639740938616|
| 139.6|             209.4|    0.0|192.0|             0.0|          1047.0|         806.9|28.0|                        28.24|[27.173735465958444]|27.173735465958444|
| 139.6|             209.4|    0.0|192.0|             0.0|          1047.0|         806.9|90.0|                        39.36| [39.92498779377876]| 39.92498779377876|
| 139.7|             163.9|  127.7|236.7|             5.8|           868.6|         655.6|28.0|                        35.23| [33.70633527711549]| 33.70633527711549|
| 140.0|             133.0|  103.0|200.0|             7.0|           916.0|         753.0|28.0|                        36.44|[32.80480524479436]| 32.80480524479436|
| 141.9|             166.6|  129.7|173.5|            10.9|           882.6|         785.3|28.0|                        44.61| [32.74609436008333]| 32.74609436008333|
| 142.0|             167.0|  130.0|174.0|            11.0|           883.0|         785.0|28.0|                        44.61|[32.88424008528651]| 32.88424008528651|
| 144.0|             136.0|  106.0|178.0|             7.0|           941.0|         774.0|28.0|                        26.14|[27.641907236052298]|27.641907236052298|
| 145.0|               0.0|  134.0|181.0|            11.0|           979.0|         812.0|28.0|                        13.21|[13.215764253351152]|13.215764253351152|
| 145.0|             116.0|  119.0|184.0|             5.7|           833.0|         800.0|28.0|                        29.16|[26.858585272574064]|26.858585272574064|
| 145.4|               0.0|  178.9|281.7|             7.8|           824.0|         868.7|28.0|                        10.54|[10.710317214544457]|10.710317214544457|

only showing top 20 rows
```

Figure 12.17 – Prediction results from the model MOJO

As you can see, we had specifically set `withDetailedPredictionCol` to `False` when loading the MOJO. That is why we can't see the detailed `_prediction_column` in the prediction results.

> **Tip**
>
> There are plenty of configurations that you can set up when loading H2O model MOJOs into Sparkling Water. There are also additional methods available for MOJO models that can help gather more information about your model MOJO. All these details can be found on H2O's official documentation page at `https://docs.h2o.ai/sparkling-water/3.2/latest-stable/doc/deployment/load_mojo.html#loading-and-usage-of-h2o-3-mojo-model`.

Congratulations – you have just learned how to use Spark and H2O AutoML together using H2O Sparkling Water.

Summary

In this chapter, we learned how to use H2O AutoML with Apache Spark using an H2O system called H2O Sparkling Water. We started by understanding what Apache Spark is. We investigated the various components that make up the Spark software. Then, we dived deeper into its architecture and understood how it uses a cluster of computers to perform data analysis. We investigated the Spark cluster manager, the Spark driver, Executor, and also the Spark Context. Then, we dived deeper into RDDs and understood how Spark uses them to perform lazy evaluations on transformation operations on the dataset. We also understood that Spark is smart enough to manage its resources efficiently and remove any unused RDDs during operations.

Building on top of this knowledge of Spark, we started exploring what H2O Sparkling Water is and how it uses Spark and H2O together in a seamlessly integrated system. We then dove deeper into its architecture and understood its two types of backends that can be used to deploy the system. We also understood how it handles data interchange between Spark and H2O.

Once we had a clear idea of what H2O Sparkling Water was, we proceeded with the practical implementation of using the system. We learned how to download and install the system and the strict dependencies it needs to run smoothly. We also explored the various configuration tunings that are recommended by H2O.ai when starting H2O Sparkling Water. Once the system was up and running, we performed an experiment where we used the Concrete Compressive Strength dataset to make predictions on the compressive strength of concrete using H2O Sparkling Water. We imported the dataset into a Spark cluster, performed AutoML using H2O AutoML, and used the leading model to make predictions. Finally, we learned how to export and import model MOJOs into H2O Sparkling Water and use them to make predictions.

In the next chapter, we shall explore a few case studies conducted by H2O.ai and understand the real-world implementation of H2O by businesses and how H2O helped them solve their ML problems.

13
Using H2O AutoML with Other Technologies

In the last few chapters, we have been exploring how we can use H2O AutoML in production. We saw how we can use H2O models as POJOs and MOJOs as portable objects that can make predictions. However, in actual production environments, you will often be using multiple technologies to meet various technical requirements. The collaboration of such technologies plays a big role in the seamless functionality of your system.

Thus, it is important to know how we can use H2O models in collaboration with other commonly used technologies in the ML domain. In this chapter, we shall explore and implement H2O with some of these technologies and see how we can build systems that can work together to provide a collaborative benefit.

First, we will investigate how we can host an H2O prediction service as a web service using the **Spring Boot** application. Then, we will explore how we can perform real-time prediction using H2O with **Apache Storm**.

In this chapter, we will cover the following topics:

- Using H2O AutoML and Spring Boot
- Using H2O AutoML and Apache Storm

By the end of this chapter, you should have a better understanding of how you can use models trained using H2O AutoML with different technologies to make predictions in different scenarios.

Technical requirements

For this chapter, you will require the following:

- The latest version of your preferred web browser.

- An **Integrated Development Environment (IDE)** of your choice.

- All the experiments conducted in this chapter have been performed using IntelliJ IDE on an Ubuntu Linux system. You are free to follow along using the same setup or perform the same experiments using IDEs and operating systems that you are comfortable with.

All code examples for this chapter can be found on GitHub at `https://github.com/ PacktPublishing/Practical-Automated-Machine-Learning-on-H2O/tree/ main/Chapter%2013`.

Let's jump right into the first section, where we'll learn how to host models trained using H2O AutoML on a web application created using Spring Boot.

Using H2O AutoML and Spring Boot

In today's times, most software services that are created are hosted on the internet, where they can be made accessible to all internet users. All of this is done using web applications hosted on web servers. Even prediction services that use ML can be made available to the public by hosting them on web applications.

The **Spring Framework** is one of the most commonly used open source web application frameworks to create websites and web applications. It is based on the Java platform and, as such, can be run on any system with a JVM. **Spring Boot** is an extension of the Spring Framework that provides a preconfigured setup for your web application out of the box. This helps you quickly set up your web application without the need to implement the underlying pipelining needed to configure and host your web service.

So, let's dive into the implementation by understanding the problem statement.

Understanding the problem statement

Let's assume you are working for a wine manufacturing company. The officials have a requirement where they want to automate the process of calculating the quality of wine and its color. The service should be available as a web service where the quality assurance executive can provide some information about the wine's attributes, and the service uses these details and an underlying ML model to predict the quality of the wine as well as its color.

So, technically, we will need two models to make the full prediction. One will be a regression model that predicts the quality of the wine, while the other will be a classification model that predicts the color of the wine.

We can use a combination of the Red Wine Quality and White Wine Quality datasets and run H2O AutoML on it to train the models. You can find the datasets at `https://archive.ics.uci.edu/ml/datasets/Wine+Quality`. The combined dataset is already present at `https://github.com/PacktPublishing/Practical-Automated-Machine-Learning-on-H2O/tree/main/Chapter%2013/h2o_spring_boot/h2o_spring_boot`.

The following screenshot shows a sample of the dataset:

fixed acidity	volatile acidity	citric acid	residual sugar	chlorides	free sulfur dioxide	total sulfur dioxide	density	pH	sulphates	alcohol	quality	color
6.8	0.18	0.37	1.6	0.055	47	154	0.9934	3.08	0.45	9.1	5	white
7.2	0.27	0.74	12.5	0.037	47	156	0.9981	3.04	0.44	8.7	5	white
7	0.39	0.31	5.3	0.169	32	162	0.9965	3.2	0.48	9.4	5	white
9.2	0.25	0.34	1.2	0.026	31	93	0.9916	2.93	0.37	11.3	7	white
7.4	0.35	0.24	6	0.042	28	123	0.99304	3.14	0.44	11.3	5	white
6.5	0.3	0.39	7.8	0.038	61	219	0.9959	3.19	0.5	9.4	5	white
7.8	0.76	0.04	2.3	0.092	15	54	0.997	3.26	0.65	9.8	5	red
...	...	...	...	...	...	...	...	...	...	...	...	...

Figure 13.1 – Wine quality and color dataset

This dataset consists of the following features:

- **fixed acidity**: This feature explains the amount of acidity that is non-volatile, meaning it does not evaporate over a certain period.

- **volatile acidity**: This feature explains the amount of acidity that is volatile, meaning it will evaporate over a certain period.

- **citric acid**: This feature explains the amount of citric acid present in the wine.

- **residual sugar**: This feature explains the amount of residual sugar present in the wine.

- **chlorides**: This feature explains the number of chlorides present in the wine.

- **free sulfur dioxide**: This feature explains the amount of free sulfur dioxide present in the wine.

- **total sulfur dioxide**: This feature explains the amount of total sulfur dioxide present in the wine.

- **density**: This feature explains the density of the wine.

- **pH**: This feature explains the pH value of the wine, with 0 being the most acidic and 14 being the most basic.

- **sulphates**: This feature explains the number of sulfates present in the wine.

- **alcohol**: This feature explains the amount of alcohol present in the wine.

- **quality**: This is the response column, which notes the quality of the wine. 0 indicates that the wine is very bad, while 10 indicates that the wine is excellent.

- **color**: This feature represents the color of the wine.

Now that we understand the problem statement and the dataset that we will be working with, let's design the architecture to show how this web service will work.

Designing the architecture

Before we dive deep into the implementation of the service, let's look at the overall architecture of how all of the technologies should work together. The following is the architecture diagram of the wine quality and color prediction web service:

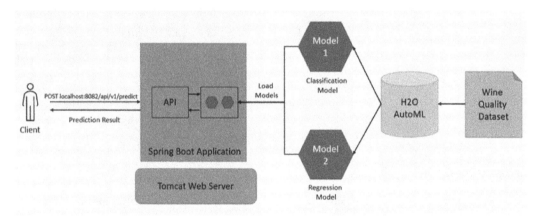

Figure 13.2 – Architecture of the wine quality and color prediction web service

Let's understand the various components of this architecture:

- **Client**: This is the person – or in this case, the wine quality assurance executive – who will be using the application. The client communicates with the web application by making a POST request to it, passing the attributes of the wine, and getting the quality and color of the wine as a prediction response.

- **Spring Boot Application**: This is the web application that runs on a web server and is responsible for performing the computation processes. In our scenario, this is the application that will be accepting the POST request from the client, feeding the data to the model, getting the prediction results, and sending the results back to the client as a response.

- **Tomcat Web server**: The web server is nothing but the software and hardware that handles the HTTP communication over the internet. For our scenario, we shall be using the Apache Tomcat web server. Apache Tomcat is a free and open source HTTP web server written in Java. The web server is responsible for forwarding client requests to the web application.

- **Models**: These are the trained models in the form of POJOs that will be loaded onto the Spring Boot application. The application will use these POJOs using the `h2o-genmodel` library to make predictions.

- **H2O server**: Models will be trained using the H2O server. As we saw in *Chapter 1, Understanding H2O AutoML Basics*, we can run H2O AutoML on an H2O server. We shall do the same for our scenario by starting an H2O server, training the models using H2O AutoML, and then downloading the trained models as POJOs so that we can load them into the Spring Boot application.

- **Dataset**: This is the wine quality dataset that we are using to train our models. As stated in the previous section, this dataset is a combination of the Red Wine Quality and White Wine Quality datasets.

Now that we have a good understanding of how we are going to create our wine quality and color prediction web service, let's move on to its implementation.

Working on the implementation

This service has already been built and is available on GitHub. The code base can be found at `https://github.com/PacktPublishing/Practical-Automated-Machine-Learning-on-H2O/tree/main/Chapter%2013/h2o_spring_boot`.

Before we dive into the code, make sure your system meets the following minimum requirements:

- Java version 8 and above

- The latest version of Maven, preferably version 3.8.6

- Python version 3.7 and above

- H2O Python library installed using pip3

- Git installed on your system

First, we will clone the GitHub repository, open it in our preferred IDE, and go through the files to understand the whole process. The following steps have been performed on *Ubuntu 22.04 LTS* and we are using **IntelliJ IDEA** *version 2022.1.4* as the IDE. Feel free to use any IDE of your choice that supports Maven and the Spring Framework for better support.

So, clone the GitHub repository and navigate to Chapter 13/h2o_spring_boot/. Then, you start your IDE and open the project. Once you have opened the project, you should get a directory structure similar to the following:

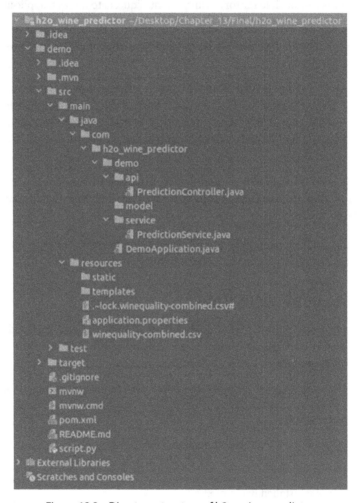

Figure 13.3 – Directory structure of h2o_wine_predictor

The directory structure consists of the following important files:

- `pom.xml`: A **Project Object Model** (**POM**) is the fundamental unit of the Maven build automation tool. It is an XML file that contains all the information about all the dependencies needed, as well as the configurations needed to correctly build the application.

- `script.py`: This is the Python script that we will use to train our models on the wine quality dataset. The script starts an H2O server instance, imports the dataset, and then runs AutoML to train the models. We shall look at it in more detail later.

- `src/main/java/com.h2o_wine_predictor.demo/api/PredictionController.java`: This is the controller file that has the request mapping to direct the POST request to execute the mapped function. The function eventually calls the actual business logic where predictions are made using the ML models and the response is sent back.

- `src/main/java/com.h2o_wine_predictor.demo/service/PredictionService.java`: This is the actual file where the business logic of making predictions resides. This function imports the POJO models and the h2o-genmodel library and uses them to predict the data received from the controller.

- `src/main/java/com.h2o_wine_predictor.demo/Demo`: This is the main function of the Spring Boot application. If you want to start the Spring Boot application, you must execute this main function, which starts the Apache Tomcat server that hosts the web application.

- `src/main/resources/winequality-combined.csv`: This is where the actual CSV dataset is stored. The Python script that trains the H2O models picks the dataset from this path and starts training the models.

You may have noticed that we don't have the model POJO files anywhere in the directory. So, let's build those. Refer to the `script.py` Python file and let's understand what is being done line by line.

The code for `script.py` is as follows:

1. The script starts by importing the dependencies:

```python
import h2o
import shutil
from h2o.automl import H2OautoML
```

2. Once importing is done, the script initializes the H2O server:

```python
h2o.init()
```

3. Once the H2O server is up and running, the script imports the dataset from the `src/main/resources` directory:

```
wine_quality_dataframe = h2o.import_file(path = "sec/
main/resources/winequality_combined.csv")
```

4. Since the column color is categorical, the script sets it to `factor`:

```
wine_quality_dataframe["color"] = wine_quality_
dataframe["color"].asfactor()
```

5. Finally, you will need a training and validation DataFrame to train and validate your model during training. Therefore, the script also splits the DataFrame into a 70/30 ratio:

```
train, valid = wine_quality_dataframe.split_
frame(ratios=[.7])
```

6. Now that the DataFrames are ready, we can begin the training process for training the first model, which is the classification model to classify the color of the wine. So, the script sets the label and features, as follows:

```
label = "color"
features = ["fixed acidity", "volatile acidity", "citric
acid", "residual sugar", "chlorides", "free sulfur
dioxide", "total sulfur dioxide", "density", "pH",
"sulphates", "alcohol"]
```

7. Now that the training data is ready, we can create the H2O AutoML object and begin the model training. The following script does this:

```
aml_for_color_predictor = H2OAutoML(max_models=10,
seed=123, exclude_algos=["StackedEnsemble"], max_runtime_
secs=300)
aml_for_color_predictor.train(x = features, y = label,
training_frame=train, validation_frame = valid)
```

When initializing the H2OautoML object, we set the exclude_algos parameter with the StackedEnsemble value. This is done as stacked ensemble models are not supported by POJOs, as we learned in *Chapter 10, Working with Plain Old Java Objects (POJOs)*.

This starts the AutoML model training process. Some print statements will help you observe the progress and results of the model training process.

8. Once the model training process is done, the script will retrieve the leader model and download it as a POJO with the correct name – that is, `WineColorPredictor` – and place it in the `tmp` directory:

```
model = aml_for_color_predictor.leader
model.model_id = "WineColorPredictor"
print(model)
model.download_pojo(path="tmp")
```

9. Next, the script will do the same for the next model – that is, the regression model – to predict the quality of the wine. It slightly tweaks the label and sets it to `quality`. The rest of the steps are the same:

```
label="quality"
aml_for_quality_predictor = H2OAutoML(max_models=10,
seed=123, exclude_algos=["StackedEnsemble"], max_runtime_
secs=300)
aml_for_quality_predictor.train(x = features, y = label,
training_frame=train, validation_frame = valid)
```

10. Once the training is finished, the script will extract the leader model, name it `WineQualityPredictor`, and download it as a POJO in the `tmp` directory:

```
model = aml_for_color_predictor.leader
model.model_id = "WineQualityPredictor"
print(model)
model.download_pojo(path="tmp")
```

11. Now that we have both model POJOs downloaded, we need to move them to the `src/main/java/com.h2o_wine_predictor.demo/model/` directory. But before we do that, we will also need to add the POJOs to the `com.h2o.wine_predictor.demo` package so that the `PredictionService.java` file can import the models. So, the script does this by creating a new file, adding the package inclusion instruction line to the file, appending the rest of the original POJO file, and saving the file in the `src/main/java/com.h2o_wine_predictor.demo/model/` directory:

```
with open("tmp/WineColorPredictor.java", "r") as raw_
model_POJO:
    with open("src/main/java/com.h2o_wine_predictor.
demo/model/ WineColorPredictor.java", "w") as model_POJO:
        model_POJO.write(f'package com.h2o_wine_
predictor.demo;\n' + raw_model_POJO.read())
```

12. It does the same for the `WineQualityPredictor` model:

```
with open("tmp/WineQualityPredictor.java", "r") as raw_
model_POJO:
    with open("src/main/java/com.h2o_wine_predictor.
demo/model/ WineQualityPredictor.java", "w") as model_
POJO:
        model_POJO.write(f'package com.h2o_wine_
predictor.demo;\n' + raw_model_POJO.read())
```

13. Finally, it deletes the `tmp` directory to clean everything up:

```
shutil.rmtree("tmp")
```

So, let's run this script and generate our models. You can do so by executing the following command in your Terminal:

```
python3 script.py
```

This should generate the respective model POJO files in the `src/main/java/com.h2o_wine_predictor.demo/model/` directory.

Now, let's observe the `PredictionService` file in the `src/main/java/com.h2o_wine_predictor.demo/service` directory.

The `PredictionService` class inside the `PredictionService` file has the following attributes:

- `wineColorPredictorModel`: This is an attribute of the `EasyPredictModelWrapper` type. It is a class from the h2o-genmodel library that is imported by the `PredictionService` file. We use this attribute to load the `WineColorPredictor` model that we just generated using `script.py`. We shall use this attribute to make predictions on the incoming request later.

- `wineQualityPredictorModel`: Similar to `wineColorPredictorModel`, this is the wine quality equivalent attribute that uses the same `EasyPredictModelWrapper`. This attribute will be used to load the `WineQualityPredictor` model and use it to make predictions on the quality of the wine.

Now that we understand the attributes of this file, let's check out the methods, which are as follows:

- `createJsonResponse()`: This function is pretty straightforward in the sense that it takes the binomial classification prediction result from the `WineColorPredictor` model and the regression prediction result from the `WineQualityPredictor` model and combines them into a JSON response that the web application sends back to the client.

- `predictColor()`: This function uses the `wineColorPredictorModel` attribute of the `PredictionService` class to make predictions on the data. It outputs the prediction result of the color of the wine as a `BinomialModelPrediction` object, which is a part of the h2o-genmodel library.

- `predictQuality()`: This function uses the `wineQualityPredictorModel` attribute of the `PredictionService` class to make predictions on the data. It outputs the prediction result of the quality of the wine as a `RegressionModelPrediction` object, which is part of the h2o-genmodel library.

- `fillRowDataFromHttpRequest()`: This function is responsible for converting the feature values received from the POST request into a `RowData` object that will be passed to `wineQualityPredictorModel` and `wineColorPredictorModel` to make predictions. `RowData` is an object from the h2o-genmodel library.

- `getPrediction()`: This is called by `PredictionController`, which passes the feature values as a map to make predictions on. This function internally calls all the previously mentioned functions and orchestrates the entire prediction process:

1. It gets the feature values from the POST request as input. It passes these values, which are in the form of `Map` objects, to `fillRowDataFromHttpRequest()`, which converts them into the `RowData` type.

2. Then, it passes this `RowData` to the `predictColor()` and `predictQuality()` functions to get the prediction values.

3. Afterward, it passes these results to the `createJsonResponse()` function to create an appropriate JSON response with the prediction values and returns the JSON to `PredictionController`, where the controller returns it to the client.

Now that we have had a chance to go through the important parts of the whole project, let's go ahead and run the application so that we can have the web service running locally on our machines. Then, we will run a simple cURL command with the wine quality feature values and see if we get the predictions as a response.

To start the application, you can do the following:

- If you are using IntelliJ IDE, then you can directly click on the green play button in the top-right corner of the IDE.

- Alternatively, you can directly run it from your command line by executing the following command inside the project directory where the pom.xml file is:

```
mvn spring-boot:run -e
```

If everything is working fine, then you should get an output similar to the following:

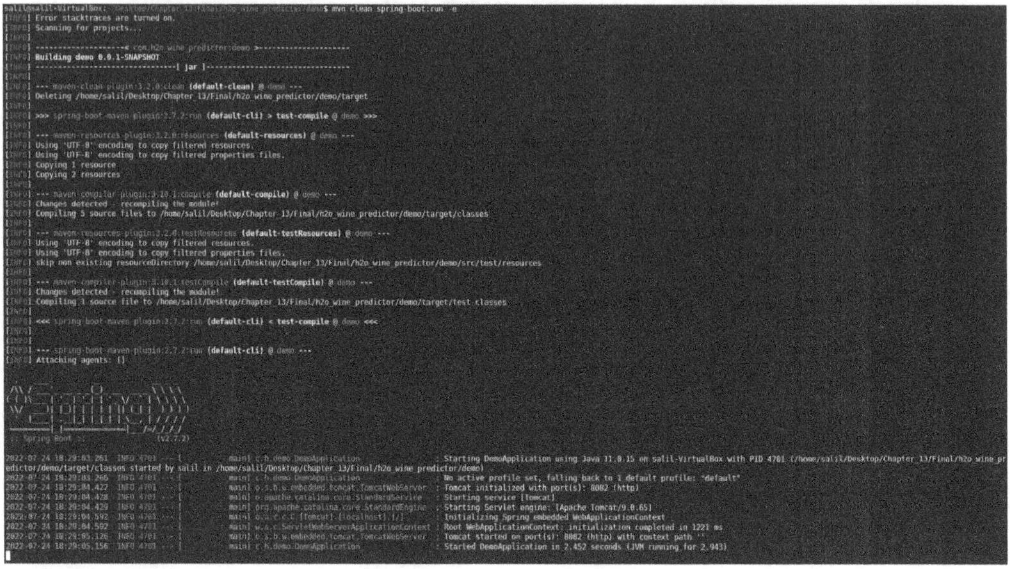

Figure 13.4 – Successful Spring Boot application run output

Now that the Spring Boot application is running, the only thing remaining is to test this out by making a POST request call to the web service running on localhost:8082.

Open another Terminal and execute the following curl command to make a prediction request:

```
curl -X POST localhost:8082/api/v1/predict -H "Content-
Type: application/json" -d '{"fixed acidity":6.8,"volatile
acidity":0.18,"citric acid":0.37,"residual
sugar":1.6,"chlorides":0.055,"free sulfur dioxide":47,"total
sulfur dioxide":154,"density":0.9934,"pH":3.08,"
,"sulphates":0.45,"alcohol":9.1}'
```

The request should go to the web application, where the application will extract the feature values, convert them into the RowData object type, pass RowData to the prediction function, get the prediction results, convert the prediction results into an appropriate JSON, and get the JSON back as a response. This should look as follows:

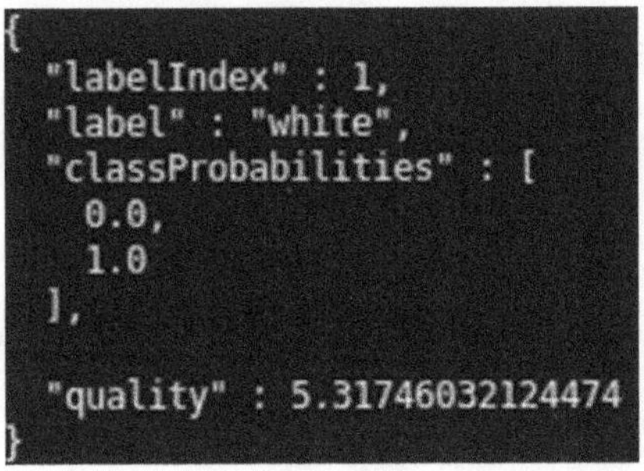

Figure 13.5 – Prediction result from the Spring Boot web application

From the JSON response, you can see that the predicted color of the wine is white and that its quality is 5.32.

Congratulations! You have just implemented an ML prediction service on a Spring Boot web application. You can further expand this service by adding a frontend that takes the feature values as input and a button that, upon being clicked, creates a POST body of all those values and sends the API request to the backend. Feel free to experiment with this project as there is plenty of scope for how you can use H2O model POJOs on a web service.

In the next section, we'll learn how to make real-time predictions using H2O AutoML, along with another interesting technology called Apache Storm.

Using H2O AutoML and Apache Storm

Apache Storm is an open source data analysis and computation tool for processing large amounts of stream data in real time. In the real world, you will often have plenty of systems that continuously generate large amounts of data. You may need to make some computations or run some processes on this data to extract useful information as it is generated in real time.

What is Apache Storm?

Let's take the example of a **log system** in a very heavily used web service. Assuming that this web service receives millions of requests per second, it is going to generate tons of logs. And you already have a system in place that stores these logs in your database. Now, this log data will eventually pile up and you will have petabytes of log data stored in your database. Querying all this historical data to process it in one go is going to be very slow and time-consuming.

What you can do is process the data as it is generated. This is where Apache Storm comes into play. You can configure your Apache Storm application to perform the needed processing and direct your log data to flow through it and then store it in your database. This will streamline the processing, making it real-time.

Apache Storm can be used for multiple use cases, such as real-time analytics, **Extract-Transform-Load** (**ETL**) data in data pipelines, and even ML. What makes Apache Storm the go-to solution for real-time processing is because of how fast it is. A benchmarking test performed by the Apache Foundation found Apache Storm to process around a million tuples per second per node. Apache Storm is also very scalable and fault-tolerant, which guarantees that it will process all the incoming real-time data.

So, let's dive deep into the architecture of Apache Storm to understand how it works.

Understanding the architecture of Apache Storm

Apache Storm uses cluster computing, similar to how **Hadoop** and even H2O work. Consider the following architectural diagram of Apache Storm:

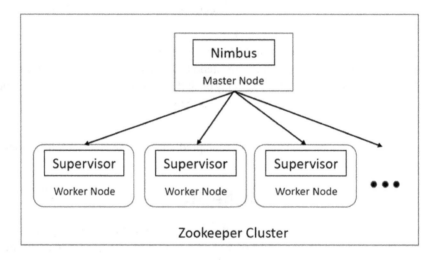

Figure 13.6 – Architecture of Apache Storm

Apache Storm distinguishes the nodes in its cluster into two categories – a master node and a worker node. The features of these nodes are as follows:

- **Master Node**: The master node runs a special daemon called **Nimbus**. The Nimbus daemon is responsible for distributing the data among all the worker nodes in the cluster. It also monitors failures and will resend the data to other nodes once a failure is detected, ensuring that no data is left out of processing.

- **Worker Node**: The worker nodes run a daemon called the **Supervisor**. The Supervisor daemon is the service that is constantly listening for work and starts or stops the underlying processes as necessary for the computation.

The communication between the master node and the worker nodes using their respective daemons is done using the **Zookeeper cluster**. In short, the Zookeeper cluster is a centralized service that maintains configuration and synchronization services for stateless groups. In this scenario, the master node and the worker nodes are stateless and fast-failing services. All the state details are stored in the Zookeeper cluster. This is beneficial as keeping the nodes stateless helps with fault tolerance as the nodes can be brought back to life and they will start working as if nothing had happened.

> **Tip**
> If you are interested in understanding the various concepts and technicalities of Zookeeper, then feel free to explore it in detail at `https://zookeeper.apache.org/`.

Before we move on to the implementation part of Apache Storm, we need to be aware of certain concepts that are important to understand how Apache Storm works. The different concepts are as follows:

- **Tuples**: Apache Storm uses a data model called Tuple as its primary unit of data that is to be processed. It is a named list of values and can be an object of any type. Apache Storm supports all primitive data types out of the box. But it can also support custom objects, which can be deserialized into primitive types.

- **Streams**: Streams are unbounded sequences of tuples. A stream represents the path from where your data flows from one transformation to the next. The basic primitives that Apache Storm provides for doing these transformations are spouts and bolts:

  - **Spouts**: A spout is a source for a stream. It is at the start of the stream from where it reads the data from the outside world. It takes this data from the outside world and sends it to a bolt.

  - **Bolt**: A bolt is a process that consumes data from single or multiple streams, transforms or processes it, and then outputs the result. You can link multiple bolts one after the other while feeding the output of one bolt as input to the next to perform complex processing. Bolts can run functions, filter data, perform aggregation, and even store data in databases. You can perform any kind of functionality you want on a bolt.

- **Topologies**: The entire orchestration of how data will be processed in real time using streams, spouts, and bolts in the form of a **Directed Acyclic Graph (DAG)** is called a **topology**. You need to submit this topology to the Nimbus daemon using the main function of Apache Storm. The topology graph contains nodes and edges, just like a regular graph structure. Each node contains processing logic and each edge shows how data is to be transferred between two nodes. Both the Nimbus and the topology are **Apache Thrift** structures, which are special type systems that allow programmers to use native types in any programming language.

> **Tip**
> You can learn more about Apache Thrift by going to `https://thrift.apache.org/docs/types`.

Now that you have a better understanding of what Apache Storm is and the various concepts involved in its implementation, we can move on to the implementation part of this section, starting with installing Apache Storm.

> **Tip**
> Apache Storm is a very powerful and sophisticated system. It has plenty of applicability outside of just machine learning and also has plenty of features and support. If you want to learn more about Apache Storm, go to `https://storm.apache.org/`.

Installing Apache Storm

Let's start by noting down the basic requirements for installing Apache Storm. They are as follows:

- Java version greater than Java 8
- The latest version of Maven, preferably version 3.8.6

So, make sure these basic requirements are already installed on your system. Now, let's start by downloading the Apache Storm repo. You can find the repo at `https://github.com/apache/storm`.

So, execute the following command to clone the repository to your system:

```
git clone https://github.com/apache/storm.git
```

Once the download is finished, you can open the `storm` folder to get a glimpse of its contents. You will notice that there are tons of files, so it can be overwhelming when you're trying to figure out where to start. Don't worry – we'll work on very simple examples that should be enough to give you a basic idea of how Apache Storm works. Then, you can branch out from there to get a better understanding of what Apache Storm has to offer.

Now, open your Terminal and navigate to the cloned repo. You will need to locally build Apache Storm itself before you can go about implementing any of the Apache Storm features. You need to do this as locally building Apache Storm generates important JAR files that get installed in your `$HOME/.m2/repository` folder. This is the folder where Maven will pick up the JAR dependencies when you build your Apache Storm application.

So, locally build Apache Storm by executing the following command at the root of the repository:

```
mvn clean install -DskipTests=true
```

The build might take some time, considering that Maven will be building several JAR files that are important dependencies to your application. So, while that is happening, let's understand the problem statement that we will be working on.

Understanding the problem statement

Let's assume you are working for a medical company. The medical officials have a requirement, where they want to create a system that predicts whether the person is likely to suffer from any complications after surviving a heart failure or whether they are safe to be discharged. The catch is that this prediction service will be used by all the hospitals in the country, and they need immediate prediction results so that the doctors can decide whether to keep the patient admitted for a few days to monitor their health or decide to discharge them.

So, the machine learning problem is that there will be streams of data that our system will need to make immediate predictions. We can set up a Apache Storm application that streams all the data into the prediction service and deploys model POJOs trained using H2O AutoML to make the predictions.

We can train the models on the Heart Failure Clinical dataset, which can be found at `https://archive.ics.uci.edu/ml/datasets/Heart+failure+clinical+records`.

The following screenshot shows some sample content from the dataset:

age	anemia	creatinine_ phosphokinase	high_ blood_ pressure	diabetes	ejection_ fraction	platelets	sex	serum_ creatinine	serum_ sodium	smoking	time	complications
75	0	582	1	0	20	265000	1	1.9	130	0	4	1
55	0	7861	0	0	38	263358	1	1.1	136	0	6	1
65	0	146	0	0	20	162000	1	1.3	129	1	7	1
50	1	111	0	0	20	210000	1	1.9	137	0	7	1
65	1	160	0	1	20	327000	0	2.7	116	0	8	1
90	1	47	1	0	40	204000	1	2.1	132	1	8	1
...	...	...	...	...	...	...	...	...	...	...	...	...

Figure 13.7 – Heart Failure Clinical dataset

This dataset consists of the following features:

- **age**: This feature indicates the age of the patient in years
- **anemia**: This feature indicates the decrease of red blood cells or hemoglobin, where 1 indicates yes and 0 indicates no
- **high blood pressure**: This feature indicates if the patient has hypertension, where 1 indicates yes and 0 indicates no
- **creatinine phosphokinase**: This feature indicates the level of the CPK enzyme in the blood in **micrograms per liter (mcg/L)**
- **diabetes**: This feature indicates if the patient has diabetes, where 1 indicates yes and 0 indicates no
- **ejection fraction**: This feature indicates the percentage of blood leaving the heart at each contraction
- **platelets**: This feature indicates the platelets in the blood in kilo platelets per **milliliter (ml)**

- **sex**: This feature indicates if the patient is a man or woman, where 1 indicates the patient is a woman and 0 indicates the patient is a man

- **serum creatinine**: This feature indicates the level of serum creatinine in the blood in **milligrams per deciliter (mg/dL)**

- **serum sodium**: This feature indicates the level of serum sodium in the blood **milliequivalent per liter (mEq/L)**

- **smoking**: This feature indicates if the patient smokes or not, where 1 indicates yes and 0 indicates no

- **time**: This feature indicates the number of follow-ups in days

- **complications**: This feature indicates if the patient faced any complications during the follow-up period, where 1 indicates yes and 0 indicates no

Now that we understand the problem statement and the dataset that we will be working with, let's design the architecture of how we can use Apache Storm and H2O AutoML to solve this problem.

Designing the architecture

Let's look at the overall architecture of how all the technologies should work together. Refer to the following architecture diagram of the heart failure complication prediction service:

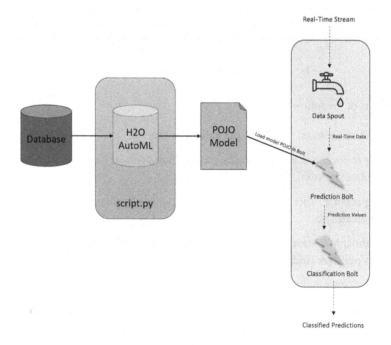

Figure 13.8 – Architecture diagram of using H2O AutoML with Apache Storm

Let's understand the various components of the architecture:

- `script.py`: From an architectural point of view, the solution is pretty simple. First, we train the models using H2O AutoML, which can be easily done by using this script, which imports the dataset, sets the label and features, and runs AutoML. The leader model can then be extracted as a POJO, which we can later use in Apache Storm to make predictions.

- **Data Spout**: We will have a spout in Apache Storm that constantly reads data and passes it to the **Prediction Bolt** in real time.

- **Prediction Bolt**: This bolt contains the prediction service that imports the trained model POJO and uses it to make predictions.

- **Classification Bolt**: The results from the Prediction Bolt are passed to this bolt. This bolt classifies the results as potential complications and no complications based on the binary classification result from the Prediction Bolt.

Now that we have designed a simple and good solution, let's move on to its implementation.

Working on the implementation

This service is already available on GitHub. The code base can be found at `https://github.com/PacktPublishing/Practical-Automated-Machine-Learning-on-H2O/tree/main/Chapter%2013/h2o_apache_storm/h2o_storm`.

So, download the repo and navigate to `/Chapter 13/h2o_apache_storm/h2o_storm/`.

You will see that we have two folders. One is the `storm-starter` directory, while the other is the `storm-streaming` directory. Let's focus on the `storm-streaming` directory first. Start your IDE and open the `storm-streaming` project. Once you open the project, you should see a directory structure similar to the following:

Figure 13.9 – storm_streaming directory structure

This directory structure consists of the following important files:

- `scripty.py`: This is the Python script that we will use to train our models on the heart failure complication dataset. The script starts an H2O server instance, imports the dataset, and then runs AutoML to train the models. We shall look at this in more detail later.

- `H2ODataSpout.java`: This is the Java file that contains the Apache Storm spout and its functionality. It reads the data from the `live_data.csv` file and forwards individual observations one at a time to the bolts, simulating the real-time flow of data.

- `H2OStormStarter.java`: This is a Java file that contains the Apache Storm topology with the two bolts – the Prediction Bolt and Classification Bolt classes. We shall start our Apache Storm service using this file.

- `training_data.csv`: This is the dataset that contains a part of the heart failure complication data that we will be using to train our models.

- `live_data.csv`: This is the dataset that contains the heart failure complication data that we will be using to simulate the real-time inflow of data into our Apache Storm application.

Unlike the previous experiments, where we made changes in a separate application repository, for this experiment, we shall make changes in Apache Storm's repository.

The following steps have been performed on *Ubuntu 22.04 LTS*; *IntelliJ IDEA version 2022.1.4* has been used as the IDE. Feel free to use any IDE of your choice that supports the Maven framework for better support.

Let's start by understanding the model training script, `script.py`. The code for `script.py` is as follows:

1. First, the script imports the dependencies:

    ```
    import h2o
    from h2o.automl import H2OautoML
    ```

2. Once importing is done, the H2O server is initialized:

    ```
    h2o.init()
    ```

3. Once the H2O server is up and running, the script imports the `training_data.csv` file:

    ```
    wine_quality_dataframe = h2o.import_file(path =
    "training_data.csv")
    ```

4. Now that the DataFrame has been imported, we can begin the training process for training the models using AutoML. So, the script sets the label and features, as follows:

```
label = "complications"
features = ["age", "anemia", "creatinine_phosphokinase",
"diabetes", "ejection_fraction", "high_blood_pressure",
"platelets", "serum_creatinine ", "serum_sodium", "sex",
"smoking", "time"]
```

5. Now, we can create the H2O AutoML object and begin the model training:

```
aml_for_complications = H2OAutoML(max_models=10,
seed=123, exclude_algos=["StackedEnsemble"], max_runtime_
secs=300)
aml_for_complications.train(x = features, y = label,
training_frame = wine_quality_dataframe )
```

Since POJOs are not supported for stacked ensemble models, we set the exclude_algos parameter with the StackedEnsemble value.

This starts the AutoML model training process. Some print statements are in here that will help you observe the progress and results of the model training process.

6. Once the model training process is done, the script retrieves the leader model and downloads it as a POJO with the correct name – that is, HeartFailureComplications – and places it in the tmp directory:

```
model = aml_for_color_predictor.leader
model.model_id = "HeartFailureComplications"
print(model)
model.download_pojo(path="tmp")
```

So, let's run this script and generate our model. Executing the following command in your Terminal:

python3 script.py

This should generate the respective model POJO files in the tmp directory.

Now, let's investigate the next file in the repository: `H2ODataSpout.java`. The `H2ODataSpout` class in the Java file has a few attributes and functions that are important for building the Apache Storm applications. We won't focus on them much, but let's have a look at the functions that do play a bigger role in the business logic of the applications. They are as follows:

- `nextTuple()`: This function contains the logic of reading the data from the `live_data.csv` file and emits the data row by row to the Prediction Bolt. Let's have a quick look at the code:

    I.   First, you have the sleep timer. Apache Storm, as we know, is a super-fast real-time data processing system. Observing our live data flowing through the system will be difficult for us, so the `sleep` function ensures that there is a delay of 1,000 milliseconds so that we can easily observe the flow of data and see the results:

    ```
    Util.sleep(1000)
    ```

    II. The function then instantiates the `live_data.csv` file into the program:

    ```
    File file = new File("live_data.csv")
    ```

    III. The code then declares the `observation` variable. This is nothing but the individual row data that will be read and stored in this variable by the spout:

    ```
    String[] observation = null;
    ```

    IV. Then, we have the logic where the spout program reads the row in the data. Which row to read is decided by the `_cnt` atomic integer, which gets incremented as the spout reads and emits the row to the Prediction Bolt in an infinite loop. This infinite loop simulates the continuous flow of data, despite `live_data.csv` containing only limited data:

    ```
    try {
            String line="";
            BufferedReader br = new BufferedReader(new
    FileReader(file));
            while (i++<=_cnt.get()) line = br.readLine(); //
    stream thru to next line
            observation = line.split(",");
        } catch (Exception e) {
            e.printStackTrace();
            _cnt.set(0);
        }
    ```

V.   Then, we have the atomic number increment so that the next iteration picks up the next row in the data:

```
_cnt.getAndIncrement();
if (_cnt.get() == 1000) _cnt.set(0);
```

VI.   Finally, we have the `_collector.emit()` function, which emits the row data so that it's stored in `_collector`, which, in turn, is consumed by the Prediction Bolt:

```
_collector.emit(new Values(observation));
```

- `declareOutputFields()`: In this method, we declare the headers of our data. We can extract and use the headers from our trained AutoML model POJO using its NAMES attribute:

```
LinkedList<String> fields_list =
new LinkedList<String>(Arrays.
asList(ComplicationPredictorModel.NAMES));
fields_list.add(0,"complication");
String[] fields = fields_list.toArray(new String[fields_
list.size()]);
declarer.declare(new Fields(fields));
```

- *Other miscellaneous functions*: The remaining `open()`, `close()`, `ack()`, `fail()`, and `getComponentConfiguration()` functions are supportive functions for error handling and preprocessing or postprocessing activities that you might want to do in the spout. To keep this experiment simple, we won't dwell on them too much.

Moving on, let's investigate the `H2OStormStarter.java` file. This file contains both bolts that are needed for performing the predictions and classification, as well as the `h2o_storm()` function, which builds the Apache Storm topology and passes it onto the Apache Storm cluster. Let's dive deep into the individual attributes:

- `class PredictionBolt`: This is the `Bolt` class and is responsible for obtaining the class probabilities of the heart failure complication dataset. It imports the H2O model POJO and uses it to calculate the class probabilities of the incoming row data. It has three functions – `prepare()`, `execute()` and `declareOutputFields()`. We shall only focus on the `execute` function since it contains the execution logic of the bolt; the rest are supportive functions. The `execute` function contains the following code:

I.   The very first thing this function does is import the H2O model POJO:

```
HeartFailureComplications h2oModel = new
HeartFailureComplications();
```

II.   Then, it extracts the input tuple values from its parameter variables and stores them in the raw_data variable:

```
ArrayList<String> stringData = new ArrayList<String>();
for (Object tuple_value : tuple.getValues()) stringData.
add((String) tuple_value);
String[] rawData = stringData.toArray(new
String[stringData.size()]);
```

III.   Next, the code categorically maps all the categorical data in the row:

```
double data[] = new double[rawData.length-1];
String[] columnName = tuple.getFields().toList().
toArray(new String[tuple.size()]);


for (int I = 1; i < rawData.length; ++i) {
    data[i-1] = h2oModel.getDomainValues(columnName[i]) ==
null
                ? Double.valueOf(rawData[i])
              : h2oModel.mapEnum(h2oModel.
getColIdx(columnName[i]), rawData[i]);
}
```

IV.   Then, the code gets the prediction and emits the results:

```
double[] predictions = new double [h2oModel.
nclasses()+1];
h2oModel.score0(data, predictions);
_collector.emit(tuple, new Values(rawData[0],
predictions[1]));
```

V.   Finally, the code acknowledges the tuple so that the spout is informed about its consumption and won't resend the tuple for retry:

```
_collector.ack(tuple);
```

- **Classifier Bolt**: This bolt receives the input from the Prediction Bolt. The functionality of this bolt is very simple: it takes the class probabilities from the input and compares them against the threshold value to determine what the predicted outcome will be. Similar to the Prediction Bolt, this `Bolt` class also has some supportive functions, along with the main `execute()` function. Let's dive deep into this to understand what is going on in the function:

  - The function simply computes if there is a possibility of *Possible Complication* or *No Complications* based on the `_threshold` value and emits the result back:

    ```
    _collector.emit(tuple, new Values(expected,
    complicationProb <= _threshold ? "No Complication" :
    "Possible Complication"));
    _collector.ack(tuple);
    ```

- `h2o_storm()`: This is the main function of the application and builds the topology using `H2ODataSpout` and the two bolts – Prediction Bolt and Classifier Bolt. Let's have a deeper look into its functionality.

  I.   First, the function instantiates `TopologyBuilder()`:

  ```
  TopologyBuilder builder = new TopologyBuilder();
  ```

  II.   Using this object, it builds the topology by setting the spout and the bolts, as follows:

  ```
  builder.setSpout("inputDataRow", new H2ODataSpout(), 10);
  builder.setBolt("scoreProbabilities", new
  PredictionBolt(), 3).shuffleGrouping("inputDataRow");
  builder.setBolt("classifyResults", new ClassifierBolt(),
  3).shuffleGrouping("scoreProbabilities");
  ```

  III.   Apache Storm also needs some configuration data to set up its cluster. Since we are creating a simple example, we can just use the default configurations, as follows:

  ```
  Config conf = new Config();
  ```

  IV.   Finally, it creates a cluster and submits the topology it created, along with the configuration:

  ```
  LocalCluster cluster = new LocalCluster();
  cluster.submitTopology("HeartComplicationPredictor",
  conf, builder.createTopology());
  ```

V.   After that, there are some functions to wrap the whole experiment together. The Util. sleep() function is used to pause for an hour so that Apache Storm can loop over the functionality indefinitely while simulating a continuous flow of real-time data. The cluster.killTopology() function kills the HeartComplicationPredictor topology, which stops the simulation in the cluster. Finally, the cluster.shutdown() function brings down the Apache Storm cluster, freeing up the resources:

```
Utils.sleep(1000 * 60 * 60);
cluster.killTopology("HeartComplicationPredictor");
cluster.shutdown();
```

Now that we have a better understanding of the contents of the files and how we are going to be running our service, let's proceed and look at the contents of the storm-starter project. The directory structure will be as follows:

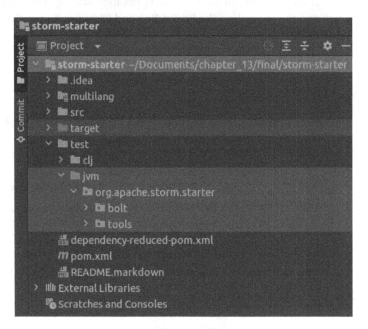

Figure 13.10 – storm-starter directory structure

The src directory contains several different types of Apache Storm topology samples that you can choose to experiment with. I highly recommend that you do so as that will help you get a better understanding of how versatile Apache Storm is when it comes to configuring your streaming service for different needs.

However, we shall perform this experiment in the test directory to keep our files isolated from the ones in the src directory. So, let's see how we can run this experiment.

Follow these steps to build and run the experiment:

1. In the `storm-streaming` directory, run the `script.py` file to generate the H2O model POJO. The script should run H2O AutoML and generate a leaderboard. The leader model will be extracted, renamed `HeartFailureComplications`, and downloaded as a POJO. Run the following command in your Terminal:

   `python3 script.py`

2. The `HeartFailureComplications` POJO will be imported by the other files in the `storm-starter` project, so to ensure that it can be correctly imported by files in the same package, we need to add this POJO to that same package. So, modify the POJO file to add the `storm.starter` package as the first line.

3. Now, move the `HeartFailureComplications` POJO file, the `H2ODataSpout.java` file, and the `H2OStormStarted.java` file inside the `storm-starter` repository inside its `storm-starter/test/jvm/org.apache.storm.starter` directory.

4. Next, we need to import the `h2o-model.jar` file into the `storm-starter` project. We can do so by adding the following dependency to the `pom.xml` file of the experiment, as follows:

   ```
   <dependency>
           <groupId>ai.h2o</groupId>
           <artifactId>h2o-genmodel</artifactId>
           <version>3.36.1.3</version>
   </dependency>
   ```

   Your directory should now look as follows:

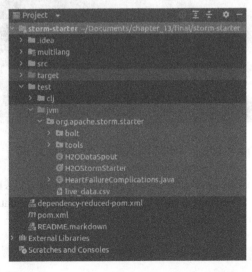

Figure 13.11 – storm-starter directory structure after file transfers

5.   Finally, we will run this project by right-clicking on the H2OStormStarter.java file and running it. You should get a stream of constant output that demonstrates your spout and bolt in action. This can be seen in the following screenshot:

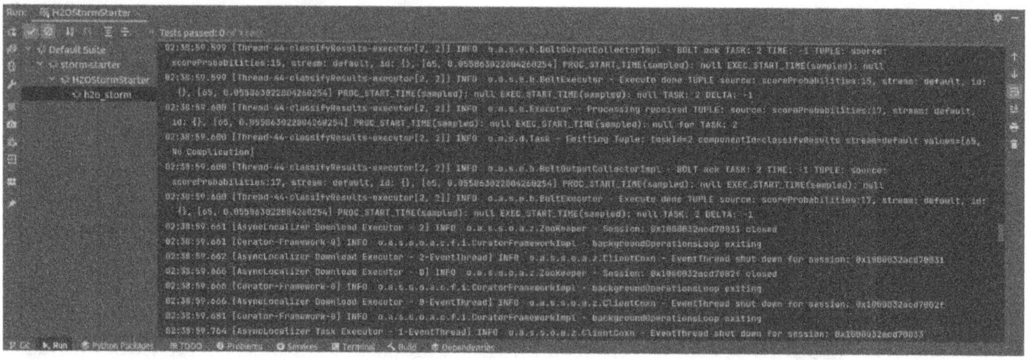

Figure 13.12 – Heart complication prediction output in Apache Storm

If you observe the results closely, you should see that there are executors in the logs; all the Apache Storm spouts and bolts are internal executor processes that run on the cluster. You will also see the prediction probabilities besides each tuple. This should look as follows:

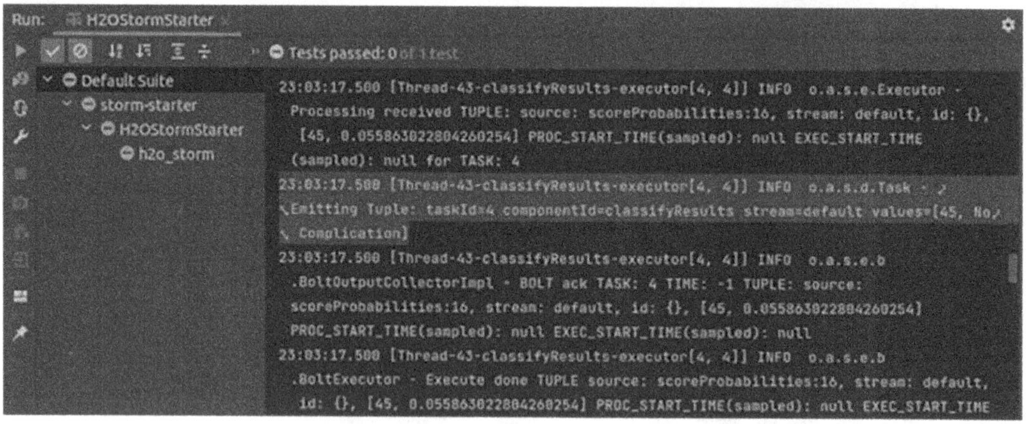

Figure 13.13 – Heart complication prediction result

Congratulations – we have just covered another design pattern that shows us how we can use models trained using H2O AutoML to make real-time predictions on streaming data using Apache Storm. This concludes the last experiment of this chapter.

Summary

In this chapter, we focused on how we can implement models that have been trained using H2O AutoML in different scenarios using different technologies to make predictions on different kinds of data.

We started by implementing an AutoML leader model in a scenario where we tried to make predictions on data over a web service. We created a simple web service that was hosted on localhost using Spring Boot and the Apache Tomcat web server. We trained the model on data using AutoML, extracted the leader model as a POJO, and loaded that POJO as a class in the web application. By doing this, the application was able to use the model to make predictions on the data that it received as a POST request, responding with the prediction results.

Then, we looked into another design pattern where we aimed to make predictions on real-time data. We had to implement a system that can simulate the real-time flow of data. We did this with Apache Storm. First, we dived deep into understanding what Apache Storm is, its architecture, and how it works by using spouts and bolts. Using this knowledge, we built a real-time data streaming application. We deployed our AutoML trained model in a Prediction Bolt where the Apache Storm application was able to use the model to make predictions on the real-time streaming data.

This concludes the final chapter of this book. There are still innumerable features, concepts, and design patterns that we can work with while using H2O AutoML. The more you experiment with this technology, the better you will get at implementing it. Thus, it is highly recommended that you keep experimenting with this technology and discover new ways of solving ML problems while automating your ML workflows.

Index

Symbols

Q

R

S

`Packt.com`

Subscribe to our online digital library for full access to over 7,000 books and videos, as well as industry leading tools to help you plan your personal development and advance your career. For more information, please visit our website.

Why subscribe?

- Spend less time learning and more time coding with practical eBooks and Videos from over 4,000 industry professionals

- Improve your learning with Skill Plans built especially for you

- Get a free eBook or video every month

- Fully searchable for easy access to vital information

- Copy and paste, print, and bookmark content

Did you know that Packt offers eBook versions of every book published, with PDF and ePub files available? You can upgrade to the eBook version at `packt.com` and as a print book customer, you are entitled to a discount on the eBook copy. Get in touch with us at `customercare@packtpub.com` for more details.

At `www.packt.com`, you can also read a collection of free technical articles, sign up for a range of free newsletters, and receive exclusive discounts and offers on Packt books and eBooks.

Other Books You May Enjoy

If you enjoyed this book, you may be interested in these other books by Packt:

Machine Learning at Scale with H2O

Gregory Keys , David Whiting

ISBN: 9781-80056-601-9

- Build and deploy machine learning models using H2O
- Explore advanced model-building techniques
- Integrate Spark and H2O code using H2O Sparkling Water
- Launch self-service model building environments
- Deploy H2O models in a variety of target systems and scoring contexts
- Expand your machine learning capabilities on the H2O AI Cloud

Machine Learning on Kubernetes

Faisal Masood , Ross Brigoli

ISBN: 9781-80324-180-7

- Understand the different stages of a machine learning project
- Use open source software to build a machine learning platform on Kubernetes
- Implement a complete ML project using the machine learning platform presented in this book
- Improve on your organization's collaborative journey toward machine learning
- Discover how to use the platform as a data engineer, ML engineer, or data scientist
- Find out how to apply machine learning to solve real business problems

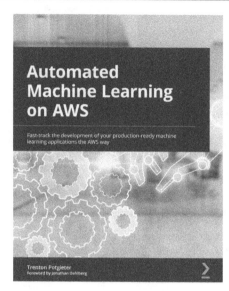

Automated Machine Learning on AWS

Trenton Potgieter

ISBN: 9781-80181-182-8

- Employ SageMaker Autopilot and Amazon SageMaker SDK to automate the machine learning process
- Understand how to use AutoGluon to automate complicated model building tasks
- Use the AWS CDK to codify the machine learning process
- Create, deploy, and rebuild a CI/CD pipeline on AWS
- Build an ML workflow using AWS Step Functions and the Data Science SDK
- Leverage the Amazon SageMaker Feature Store to automate the machine
- learning software development life cycle (MLSDLC)
- Discover how to use Amazon MWAA for a data-centric ML process

Packt is searching for authors like you

If you're interested in becoming an author for Packt, please visit `authors.packtpub.com` and apply today. We have worked with thousands of developers and tech professionals, just like you, to help them share their insight with the global tech community. You can make a general application, apply for a specific hot topic that we are recruiting an author for, or submit your own idea.

Share Your Thoughts

Now you've finished *Practical Automated Machine Learning Using H2O.ai*, we'd love to hear your thoughts! Scan the QR code below to go straight to the Amazon review page for this book and share your feedback or leave a review on the site that you purchased it from.

`https://packt.link/r/1-801-07452-6`

Your review is important to us and the tech community and will help us make sure we're delivering excellent quality content.